EUTHANASIA IS *NOT* THE ANSWER

EUTHANASIA
IS
NOT
THE
ANSWER

A Hospice Physician's View

by David Cundiff, MD

✳ Humana Press • Totowa, New Jersey

To

My Patients—
past, present, and future

Library of Congress CIP information:

Cundiff, David
        Euthanasia is not the answer: a hospice physician's view/ by
    David Cundiff
            190 + viii pp., 15.24 x 22.86 cm
        ISBN 0-89603-237-X
            1. Euthanasia 2. Hospice Care. I. Title
    R726.C86 1992
    179'.7—dc20   92-17245
                                                              CIP

10 9 8 7 6 5 4 3 2 1

CONTENTS

Preface

Instances of euthanasia or mercy killing date back to antiquity. However, it is only recently that the unprecedented grassroots efforts to legalize euthanasia have begun building. "Terminal Illness, Assistance with Dying," a California ballot initiative for the November 1992 election, might for the first time in modern history legalize euthanasia and assisted suicide by physicians. Similar initiatives are planned in other states. To vote intelligently, citizens in California and throughout the United States need to learn who is likely to request euthanasia or assisted suicide, and why.

How we care for the terminally ill eventually affects us all. In over half of all deaths, a chronic disease process such as cancer or congestive heart failure leads to a terminal phase that may last for days, weeks, or months. Most people are more afraid of the suffering associated with this terminal phase than they are afraid of dying itself. When polled, most Americans tell us they would prefer to die at home, surrounded by loved ones, rather than in a hospital receiving high-tech tests and treatments until the last. Yet the majority of people, even those with terminal illnesses, die in the hospital. What factors in our culture and health care system have led to this dichotomy?

Unrelieved suffering is also the primary reason for euthanasia requests. But what is the nature of that unrelieved suffering? Does our medical care system actually increase suffering in many cases? What is done by the medical profession to relieve the suffering of the terminally ill? What more can be done? What barriers are there to improving the care and relieving the suffering of the terminally ill? We need to look deeply for the answers to these questions before agreeing to euthanasia as a solution to suffering.

In my view, improved care of the terminally ill will make the question of euthanasia and assisted suicide moot. But no matter whether "Terminal Illness, Assistance with Dying" passes or fails at the polls, the issue of how best to care for the dying will not go away until some fundamental changes occur in the practice of medicine in Western society. One may hope that the emergence of legalized euthanasia as a burning issue will stimulate the changes required to bring about more humane care of the terminally ill. This process of change requires that not only physicians, nurses, other health care professionals, but also the public become more informed about treatment of the variety of symptoms of terminal diseases: physical, psychological, emotional, as well as spiritual. Structural changes in our health care system itself are also required, including major alterations in the financial incentives built into the system.

You owe it to yourself—and to the whole of your society to analyze this subject, and to make up your own mind. This book is intended to help you accomplish just that. If I have succeeded, you will at the conclusion of your analysis, be prepared to act and vote in the best interest of the terminally ill—of all of us, really.

Acknowledgments

Years of experience caring for patients prepared me to write this book. My terminally ill patients with cancer and AIDS and their loved ones taught me a great deal. I learned about living, dying, feeling, suffering, respecting, loving, and caring for other people and for myself.

My thanks go to Kathy McCarthy, a superb nurse, who served with me for over four years on the Cancer Pain Service at LA County–USC Medical Center. She helped me also by editing the preview edition of this book. Cadena Bedney, the Cancer Pain Service nurse specialist, worked extra hard to save me time for this book. I also owe my thanks to many other nurses, including Susan Muratta, Diane McNamara-Margraves, Maxine Patrick, Doreen Young, Irma Aquel, Mary Nash, Sandra Williams, Olivia Escobosa, Fran Zonfrillo, and Pat Brehnerkemper.

When I was a senior in college, Ira Spicer suggested that I consider becoming a physician. The idea had never occurred to me. I never had a chance to thank him earlier, so I'll take the opportunity now. If you read this Ira, give me a call.

Sheldon Katz from Graphic Solutions in Long Beach gave rapid and expert publishing assistance to the preview edition of this book. Thomas Lanigan and his staff at Humana Press provided the editing and publishing experience necessary for these ideas to reach a wide audience.

In England, many hospice professionals graciously shared their expertise with me. Drs. Robert Twycross and Richard Lamerton were especially generous with their time and knowledge. They started me on the road to becoming a palliative care physician.

Over the past 13 years, several writers have helped me by editing and critiquing my manuscripts. These include Julie Hatoff, Neil Heiman, Harold Glicken, and Debra Arrington. Walt Murray has been especially helpful over the years, sharing his insights about health care and writing.

I appreciate the support of my wife, Jennifer, and children—Amanda, Molly, and Chelsea. My apologies to Amanda, who has chided me for paying more attention to "youth in Asia" that to youth in Long Beach.

I greatly appreciate the encouragement of Gurumayi Chidvilasananda, the current head of Siddha Yoga, who bestowed her blessings on this work. My special thanks also goes to Baba Muktananda, who in 1981 encouraged me to continue writing in order to explain cancer to the public.

CHAPTER 1

What Euthanasia Is
—What Hospice Is

The word "euthanasia" comes from a Greek phrase meaning "good death." In today's society it means killing a terminally ill person as a way to end that person's pain and suffering. Unfortunately, most people equate terminal cancer or AIDS with constant, unrelieved pain and suffering. Fortunately, today's medicine can greatly alleviate the pain and suffering from these and other diseases in all dying people. But it is also regrettably true that the majority of physicians in the United States have never been taught the techniques of treating the physical, psychological, and emotional symptoms of terminal disease.

In the United States—as in every country in the world—euthanasia is considered murder. Only in the Netherlands can physicians perform euthanasia, in certain carefully delineated circumstances, owing to a precedent set by the Dutch Supreme Court. But the Dutch legislature has still not officially legalized the practice.

In recent years there has been a growing movement toward the legalization of euthanasia in the United States. In a historic election held on November 5, 1991, voters in Washington state decided not to legalize euthanasia for consenting, terminally ill patients by voting down Initiative 119. This Initiative would have allowed physicians to carry out euthanasia by administering lethal drug overdoses to those of their terminally ill patients who requested it.

Assisting suicide means to provide a person who plans to kill himself with the means to do so. This may be accomplished by supplying a lethal overdose of medication, by providing a gun, or by other means. Physician-assisted suicide would also have been legalized by Initiative 119.

Currently in the US, except in Michigan, assisting suicide is also a crime. Michigan legislators are revising the law in the wake of the deaths of Janet Adkins in 1990 and Marjorie Wentz and Sherry Miller (1991) in Dr. Jack Kevorkian's "suicide machine."

ALLOWING A TERMINALLY ILL PERSON TO DIE NATURALLY

Euthanasia or "active euthanasia" is often confused with allowing the terminally ill person to die naturally of the disease. Allowing an individual to die means foregoing or stopping medical treatments intended to prolong life. For example, a terminally ill person on a respirator (breathing machine) in an intensive care ward may request that the machine be turned off and that he or she be allowed to die. The discontinuation of life support technol-

* * * * *

In [passive euthanasia] the person
dies naturally of the disease process;
in [active euthanasia] the person is killed.

* * * * *

ogy when any realistic hope for recovery has completely vanished is a legal, ethical, and appropriate act also known as passive euthanasia.

A poor substitute term for "allowing to die," "passive euthanasia" implies that there is a strong similarity with active euthanasia. Proponents of active euthanasia argue that the difference between passive and active euthanasia is little more than semantic. But though it is simple, the difference is much more than that. In one case, the person dies naturally of the disease process, whereas in the other, the person is killed by the injection of an overdose of medication. The US courts and medical associations also make this critical distinction.

The decision to allow a terminally ill person to die usually comes after multiple treatments have failed to cure or control a patient's disease and the prognosis is poor. Patients themselves, and family members, begin to appreciate that further heroic treatment would only prolong suffering and dying and not give a realistic chance at remission or recovery. These situations occur every day in American hospitals.

Physicians may also forego initiating life support treatment after receiving the informed consent of a terminally ill patient or proxy. Legal avenues permitting patients to forego or stop treatments that only prolong suffering include natural death acts in some states and what are known as durable power of attorney laws. These afford terminally ill patients the legal right to die without being forced to accept life-prolonging therapies.

People with cancer, AIDS, or other medical conditions are always legally free to choose whether to have particular medical treatments. No one is obligated by law to accept intravenous hydration, antibiotics for infections, renal dialysis (kidney machine) treatments for kidney failure, or other medical therapies if these would only prolong needless suffering. At some point such measures become useless in every patient with advanced incurable cancer or AIDS. For many terminally ill people, life- prolonging treatments can all too easily become a fate worse than death. And sad as it may seem, nature should be allowed to take its course at some point.

STATE EUTHANASIA INITIATIVES

The National Hemlock Society (NHS) and its political arms, the Washington Hemlock PAC, Washington Citizens for Death with Dignity, and Americans Against Human Suffering lead the campaign to legalize euthanasia. They sponsored Initiative 119 on the Washington State ballot which was defeated on November 5, 1991. NHS-sponsored "Terminal Illness. Assistance with Dying" will be on

* * * * *

Traditionally, death with dignity has meant that the dying are kept physically comfortable and given psychological, emotional, and spiritual support...

* * * * *

the California Ballot in November of 1992. Another initiative is planned in Oregon in 1994.

Traditionally, death with dignity has meant that the dying are kept physically comfortable and given psychological, emotional, and spiritual support by skilled medical professionals in conjunction with their caring families and loved ones. Meddlesome and unnecessary diagnostic procedures and therapies should be avoided and, if possible, the dying should live their last days in their own homes in their own beds. Personally, I would hardly consider it 'death with dignity' if the pain and other physical and psychological symptoms of my own terminally ill patients were so out of control that they were begging for euthanasia or assisted suicide.

The Roper Poll in May 1991 showed that people in Washington, Oregon, and California favored legalizing voluntary euthanasia for terminally ill patients almost 3 to 1. A coalition of hospice and religious groups mounted an effective campaign to educate Washington state voters about the risks of allowing euthanasia or physician-assisted suicide. Initiative 119 lost, with 54% opposed to 46% in

* * * * *

*Personally, I would hardly consider
it death with dignity if the pain
and other physical and psychological
symptoms of my own terminally ill
patients were so out of control that
they were begging for euthanasia.*

* * * * *

favor. In March, 1992 a poll showed that Californians favor the euthanasia initiative by 75% to 25%. Fewer than 1% were undecided.

Much more education of Americans about euthanasia remains to be done. Many do not understand that allowing terminally ill people to die naturally is legal. Also, few Americans realize that much better control of pain and other physical and psychological symptoms of the dying can be achieved, but often is not, because of the lack of physician and nurse training in this area.

HOSPICE OR PALLIATIVE CARE

The hospice approach to the treatment of the terminally ill focuses on relieving the physical symptoms of patients and on providing psychological and social support for both patient and family. Whereas standard medical treatment for cancer and AIDS patients strives to prolong life at virtually any cost, hospice seeks to optimize the quality of life of the patient's remaining time. The National Hospice Organization defines the hospice philosophy as:

> Hospice affirms life. Hospice exists to provide support and care for persons in the last phases of incurable disease so that they might live as fully and comfortably as possible. Hospice recognizes dying as a normal process whether or not resulting from disease. Hospice neither hastens nor postpones death. Hospice exists in the hope and belief that, through appropriate care and the promotion of a caring community sensitive to their needs, patients and families may be free to attain a degree of mental and spiritual preparation for death that is satisfactory to them.[1]

Palliative care is a synonym for hospice.

With notable exceptions, hospice or palliative care services in the United States are woefully inadequate. Improved training in hospice for all medical professionals and the allocation of a greater proportion of cancer and AIDS treatment resources to hospice care are urgently needed.

WHO WANTS EUTHANASIA AND WHY WOULD ONE WANT IT

Informal polls among cancer specialists show that requests for euthanasia or assisted suicide are very uncommon. Two of my oncologist colleagues with more than 25 years of combined experience

* * * * *

...terminally ill people wanting euthanasia or suicide cite pain as the chief factor driving them to want to end their lives.

* * * * *

reported that only two of their patients had ever asked for euthanasia or assisted suicide. The published literature confirms my impression about the rarity of euthanasia requests despite the frequency of poorly treated physical and psychological symptoms.

Why do some terminally ill people want euthanasia or assisted suicide? By all accounts terminally ill people wanting euthanasia or suicide cite *pain* as the chief factor driving them to end their lives.

I have treated several thousand cancer and AIDS patients in the past 18 years, first as a medical oncologist (cancer treatment specialist), and then as a palliative care doctor. Ten of my patients have asked for euthanasia. Another 15 or 20 attempted suicide. Only three that I know about actually succeeded in committing suicide.

In my experience the cases in which terminally ill people either requested euthanasia or committed suicide are similar. Poor pain control, other physical symptoms out of control, or inadequate psychosocial support occur invariably. Cases from the literature in which detailed information is provided confirm this impression.

* * * * *

*It is a disgrace that the majority
of our health care providers
lack the knowledge and the skills
properly to treat pain and other symptoms
of terminal disease.*

* * * * *

Tragically, in the vast majority of these cases, the pain could have been readily alleviated and other physical symptoms suffered by these patients could have been better controlled if the caregivers had expertise in palliative care techniques. More appropriate psychological, social, and spiritual support might well have been provided if the physicians, nurses, and other health care workers were adequately trained.

It is a disgrace that the majority of our health care providers lack the knowledge and the skills to treat pain and other symptoms of terminal disease properly. The absence of palliative care training for medical professionals results in suboptimal care for almost all terminally ill patients and elicits the wish to hasten their own deaths in a few.

BARRIERS TO GOOD PAIN CONTROL

In 1979, after four years of fellowship training in medical oncology and hematology, I found myself on a tour of English hospices. Frankly, a visit to England was more intriguing to me at the time than more training in cancer therapy. After all, I had just completed a four-year study of cancer diagnosis and treatment in the United States and Canada—what more could I learn?

A lot. I learned that *I didn't know **that I didn't know** how to control the pain from cancer.* Studying in the English hospices opened my eyes to my deficiencies in managing pain and other symptoms of advanced cancer. I also found that the treatment approaches for advanced cancer patients of American and British physicians are quite different.

Compared to American physicians, British doctors are much more conservative about continuing diagnostic procedures, chemotherapy, radiation, and surgeries in their cancer patients with advanced disease. Being able to experience the different approaches firsthand made me aware of the level of suffering that overly aggressive anticancer therapies often produce for patients in the US.

Most of the general public believes that physicians and nurses do all that is humanly possible to control pain from cancer. Few people realize that most physicians should be much better trained to treat the physical and psychological symptoms associated with terminal illness than they are. As a result, many people develop a sense of hopelessness, thinking that little can be done to relieve the pain and suffering of the dying process.

Most practicing physicians in the United States have not seen first-rate cancer pain management and the optimal control of the physical symptoms of cancer and AIDS. Like me before my hospice tour, *they don't know that they don't know how to do it.* Unfortunately, the norm in caring for patients with end-stage disease is to expect poor pain control with poor palliative care overall. Therefore, physicians in training imagine that when the attack on cancer is stopped there is nothing else that can be done. They think that's simply the way it is when you're dying of cancer.

Unfortunately, very few hospice training programs exist in the United States. Since most medical and nursing schools don't teach palliative care, there are inherent difficulties in attempting to educate health professionals about it later on. This lack of firsthand experience with good palliative care makes it much harder to teach doctors and nurses.

Besides an overwhelming lack of physician and nurse training in pain and symptom management, several other barriers may impede good pain control. One of these is poor communication. For a variety of reasons, patients often do not tell their physicians about their pains. In other instances, nurses may unintentionally sabotage efforts to relieve pain by conservative interpretation of doctors' orders or by delaying in giving pain medicines. Family members may obstruct the delivery of adequate pain treatment by not understanding the purpose or the schedule of the medicine. In some situations, lack of money for medications or other treatments may interfere with pain treatment.

Based on my clinical experience, information in the medical literature, and discussions with countless other doctors, I share the feeling of other specialists in hospice and pain management everywhere that most of the pain of advanced cancer and AIDS can be controlled.

RESUSCITATION

Unfortunately, many patients with advanced cancer are needlessly resuscitated and placed on life support systems when there is no reasonable hope of recovery. This is simply bad medicine. Fears of legal actions have complicated this sensitive issue. Oftentimes, physicians fail to discuss the disease prognosis as well as what to do should the patient's heart stop beating.

On the other hand, when physicians do discuss these matters, they may offer the patient a choice between doing everything, including cardiopulmonary resuscitation, and just giving up and doing nothing. These radical alternatives frustrate and frighten the patient or the next of kin, and constitute a classic "no win situation." For instance, if the son of a dying, comatose mother chooses full cardiopulmonary resuscitation and breathing machines, he will feel guilty after his mother has a few additional miserable hours or days in an intensive care unit before dying. If the son chooses to give up, forgoing resuscitation, he will wonder whether his mother would have survived had everything possible been done.

* * * * *

*...Many patients with advanced cancer are
needlessly resuscitated and placed on life
support systems when there
is no reasonable hope of recovery.
This is simply bad medicine.*

* * * * *

Physicians with good training in palliative care do not offer patients or families this impossible "Hobson's choice." These physicians begin discussing the prognosis, the resuscitation issue, and options for hospice care early in the disease process. Doctors who understand palliative care generally do not recommend resuscitation and breathing machines in intensive care units. Instead, they offer their time to care for the patient and to treat any symptoms or problems that may arise.

ENGLISH VERSUS AMERICAN HEALTH CARE SYSTEMS

Ironically, the socialized medicine environment in England seems to foster good cancer pain control and overall palliative care. The reasons for this are complex, involving the superior training of English physicians in palliative care, the more realistic expectations of the English public concerning medical technology, and incentives in the English medical system favoring palliative care. As a result of the interplay of all these factors, English general practitioners and oncologists (cancer specialists) refer terminally ill cancer and AIDS patients to hospice programs much earlier in the disease process than do US physicians.

All of this is made possible because hospice training is more readily available in England than in the United States. English physicians and nurses may enroll in numerous hospice training programs to learn about pain and symptom management. As a consequence of freely available hospice training, experts in hospice are much more available and hospital and community resources for hospice treatment are part of standard care in England.

The American medical system—or patchwork of systems—unintentionally fosters barriers to optimal pain and symptom management for terminally ill people. Lack of physician and nurse training in palliative care is a major barrier previously discussed.

Ironically, financial disincentives also discourage good pain and symptom control for doctors and for hospitals. Patients with poorly controlled pain and other symptoms fill empty hospital beds and require many more costly physician services. A good measure of the effectiveness and quality of palliative care services is the degree of success in managing the patient's symptoms at home so that only minimal time in the hospital is necessary. Consequently, the better the palliative care provided, the more money the doctor and hospital lose compared to standard oncology care.

Medical care reform needs to address the perverse disincentives that often obstruct good pain and symptom control. Increased reimbursement for Medicare and MediCaid hospice benefits tops the list of reforms needed to improve health care in America.

BETTER PALLIATIVE CARE— THE ALTERNATIVE TO EUTHANASIA

In this book I argue against the legalization of euthanasia proposed in various state ballot initiatives. However, I do not advocate the status quo, which all too frequently is uncontrolled pain and suffering for the terminally ill.

The unnecessary physical and mental torment of dying in a standard medical setting can be incredible. However, with excellent palliative care, the dying process can instead be associated with profound emotional and spiritual growth for the patients, as well as for the loved ones and caregivers.

Ideally, the debate surrounding the legalization of euthanasia should center around the inadequacies of palliative care in this country, but this has not been the case. Neither the pro-euthanasia nor the anti-euthanasia forces have sufficiently highlighted the inadequacies of palliative care training or the meager medical resources directed toward hospice. Nor has either camp offered tangible proposals for alleviating the unnecessary suffering of the terminally ill.

* * * * *

Hospice should be at the top of the agenda for health care reform...Improved hospice services can simultaneously improve the quality of care and reduce the cost.

* * * * *

Hospice should be at the top of the agenda for health care reform in this country. Improved hospice services can simultaneously improve the quality of care and reduce its cost. Increased access to medical care services naturally follows the discovery of a low-cost, high-quality alternative therapy.

Education of the public about these issues is necessary to effect change. One hopes that the controversy so strongly stirred by the state euthanasia initiatives will stimulate concerned citizens to learn much more about hospice. With increased public awareness, leaders in the health care field will follow. Action by American legislators, insurance companies, medical administrators, health educators, and other concerned citizens is essential if the hospice approach is to grow, and thus produce a dramatic improvement in the treatment of cancer patients.

My thesis in this book is that vastly improved hospice training for health care professionals, along with better quality and greater availability of hospice services can render the issue of euthanasia and assisted suicide essentially moot.

CHAPTER 2

Why—and How Often— Do Terminally Ill People Request Euthanasia?

MR. LEE—A CANCER PATIENT WHO REQUESTED EUTHANASIA

Mr. Lee (not his real name) was a 52-year-old Korean man who had undergone surgery for stomach cancer 13 months earlier. Since the cancer had already spread to the liver and elsewhere, he had been given intraoperative chemotherapy followed by conventional outpatient chemotherapy.

When the chemotherapy failed to control the disease, he received an experimental chemotherapy for six months at my hospital. This also failed to control his disease; it lowered his blood platelet count, thus increasing his chances of bleeding from the remaining abdominal tumors. As an outpatient, multiple transfusions of blood were given because of hemorrhage through the gastrointestinal tract.

One bleeding episode required hospitalization to achieve control. During that time, after discussion with his doctor, Mr. Lee agreed to a "Do Not

Resuscitate" order that was then recorded in his chart. Unfortunately, on discharge from hospital he was not referred to our visiting nurse associaton hospice program

He had been out of the hospital only three weeks when he began to vomit blood and was again rushed to our emergency room. He was immediately transfused with blood and quickly moved to our new, ultramodern intensive care unit. When bleeding persisted, he underwent angiography (an X-ray dye study) of his abdominal blood vessels and, once located, the bleeding artery was blocked off by an injection of a special material. Soon the specialists in interventional radiology repeated this procedure because of recurrent bleeding. Because the cancer was so advanced and the patient's inability to take food while the acute bleeding problem persisted, the intensive care unit physicians ordered total parenteral (intravenous) nutrition to prevent malnourishment. This provided about 3000 calories per day, along with plenty of intravenous fluid. A previously inserted catheter carried the nutritional fluids directly into the right side of his heart.

After more time in the intensive care unit, Mr. Lee developed a fever; his doctors promptly ordered antibiotics. Later, when the fever persisted and blood cultures showed infection with resistant bacteria, he was switched to more powerful antibiotics.

On the tenth hospital day, a new intensive care unit doctor discussed with Mr. Lee and his family the seriousness of his condition. Mr. Lee again requested not to be resuscitated if his heart stopped beating. The doctor dutifully noted this in the chart.

Abdominal pain had been a big problem even before this hospitalization. Mr. Lee's oncologist had prescribed prolonged-release morphine beginning at least six months before hospitalization. While in intensive care, his pain had increased despite institution of intravenous morphine infusion and titrating the dose to 20 milligrams per hour (a high dose).

Mr. Lee's doctor asked the anesthesiology pain service to give a nerve block to better control his severe pain. After deliberation for several days, the anesthesiologists declined to carry out the nerve block procedure for fear of causing internal bleeding, and possibly shortening his life.

At his wit's end, on the 21st day in the intensive care unit, the intern called on me to offer new suggestions for the management of Mr. Lee's pain. He had observed that a marked accumulation of fluid in Mr. Lee's abdomen was now also contributing to the pain.

I am a consultant to other physicians at the hospital in the management of their patients' pain from cancer and AIDS. A nurse works with me to help control that pain. Our care for terminally ill patients is rooted in the hospice philosophy and employs hospice techniques. When no cure of these patients is possible, the hospice approach emphasizes the management of pain and the provision of psychological, social, and spiritual support for both patients and family members. Using a team approach, we often call in physical therapists, psychiatrists, anesthesiologists, and other specialists.

The intern told me that because of his pain and the overall poor prognosis, Mr. Lee had been begging for a lethal overdose of medication—in short, begging for euthanasia. The young doctor was obviously in an uncomfortable position, as any doctor would be for that matter.

This case offered me an excellent opportunity to teach the intern some of the basics of palliative care. I explained that, in a terminally ill patient in this situation, although we cannot honor a request for euthanasia, physicians are under no legal, moral, or other obligation to continue therapies designed to prolong life, such as blood product transfusions, total parenteral nutrition, and antibiotics.

I suggested that a paracentesis (removal of abdominal fluid) be done to decrease the pressure in Mr. Lee's abdomen. I also requested that the intravenous fluids, including the total parenteral nutrition, be stopped to prevent further misery from the accumulation of abdominal fluid. Finally, I recommended an increase in the morphine infusion dose to 30 milligrams per hour.

The next day when I saw Mr. Lee, he had been transferred to a "closely monitored area" on a regular medical ward. He was in coma and the morphine infusion had been stopped. Very distraught relatives filed in and out of his room for short visits, making their way between the hospital staff and the life-support technology.

Skimming the chart (three volumes had accumulated during the 22-day ICU stay), I noted that the paracentesis had not been done, again for fear of causing bleeding and the shortening of Mr. Lee's life. Two expensive intravenous antibiotics, total parenteral nutrition feedings, and frequent insulin injections had continued. Blood cultures drawn two or three days earlier showed that two types of bacteria were growing despite the antibiotics. Other laboratory tests continued to be ordered. I spoke at length with the new intern and resident about what to do if pain reemerged and in general concerning palliative care in this type of situation.

During the following night, Mr. Lee woke up enough to express pain. More morphine was given intravenously, but initially did not work. Instead of giving Mr. Lee higher doses of morphine, the doctors had injected valium, which only quieted him down.

In the morning, the staff had suddenly become concerned with inappropriate utilization of the hospital's resources (the closely monitored unit) and ordered Mr. Lee's transfer to the regular ward. The total parenteral nutrition, antibiotics, and insulin could all be continued on the regular ward, but the morphine infusion pump could not.

The new intern wrote an order for prolonged-release morphine sulfate to be crushed and given through the gastric feeding tube. I pointed out to the staff the problems with this strategy. Crushing prolonged-release morphine converts it into immediate-release morphine. In someone with cancer and tense fluid throughout his abdomen, someone acutely ill with sepsis and low blood pressure, oral anagesics or other medications would not be reliably absorbed from the gastrointestinal tract. The nursing administration then made an exception and allowed the morphine pump for his final hours of life. The pump was not turned on since he never came out of coma.

I wish there was a happy ending to this story, but there is not. When Mr. Lee died, the doctors could truthfully tell the family that they had done all that could be done medically to save him. No one could be charged with malpractice since this is not unusual care of the dying in America. However, a lack of training in palliative care and the obstacles built into our medical care system had prevented even rudimentary pain and symptom control measures for Mr. Lee, let alone help with the psychological and emotional process of preparing for his death.

In a time of dire shortages of health care funding for the poor, this hospitalization cost the taxpayer over $50,000. This hospitalization served only to magnify his pain and suffering enough for him to beg for euthanasia. For Mr. Lee, euthanasia is not the answer. Physician training in palliative care offers the prospect of a far better solution.

WHY WOULD ANYONE WANT EUTHANASIA?

Uncontrolled pain and suffering heads the list of reasons for requesting euthanasia. Dying people

* * * * *

*Uncontrolled pain and suffering
heads the list of reasons
for requesting euthanasia.*

* * * * *

also fear the loss of body function, detest dependence on others, and generally dislike the sense of being a burden. This naturally distresses the patient's friends and family, who feel helpless and unsure how to respond.

If the pain or physical symptoms become intolerable, or if the burdens of 24-hour care of the patient exhaust the family, the only alternative seems to be hospitalization. Unfortunately, the technology intensive approach of the acute care hospital often only increases, rather than decreases, the patient's suffering and the family's distress. If admission to hospital is chosen, the hospital staff may sometimes make the dying patient feel unwelcome and cause the family members to feel guilty about being unable to manage the care at home.

INFREQUENCY OF REQUESTS FOR EUTHANASIA OR ASSISTED SUICIDE

Since I run a Cancer Pain Consultation Service at a busy county hospital for indigent patients, I am more likely to encounter requests for euthanasia or assisted suicide than are other physicians. However, Mr. Lee was one of only ten of my patients who have asked for euthanasia or assisted suicide.

Other patients who had requested euthanasia stopped asking for hastened death once their pain was adequately treated and their physical and emotional needs were addressed. One of my patients continued to ask for euthanasia until he died from his uncontrolled pain. Because of a lack of adequate resources to provide the palliative care he needed, he was not given a pain-relieving procedure that almost certainly would have controlled his suffering.

My experience parallels that of other oncologists and specialists in hospice medicine. Hospice physicians, nurses, and other professionals rarely encounter terminally ill people who wish to have assisted suicide or euthanasia. When it does happen, they almost always change their minds once their physical symptoms are controlled and they are placed in a caring, supportive, hospice environment.

SUICIDE IN CANCER PATIENTS

We can get an idea of the potential demand for euthanasia and assisted suicide by looking at the rate at which cancer patients now commit suicide. When we compare the rate of such patients with that of the general population, we can see approximately how important the cancer was in the person's decision to kill him- or herself. Table 1 details some representative larger studies in the area.[1-5]

These studies show that only one or two out of 1000 cancer patients per year commit suicide and

TABLE 1

STUDIES ON INCIDENCE OF SUICIDE AMONG CANCER PATIENTS

	Total suicides in cancer patients	Total cancer deaths	Risk relative to general population (1.0, average risk)	
			Men	Women
Louhivuori (1979/Finland)	63	28,857	1.3	1.9
Fox (1982/US)	192	144,530	2.3	0.9
Bolund (1985)	22	19,000	---	---

that this is approximately twice the national suicide rate of the general population.

The factors associated with an increased risk of suicide in cancer patients have been reported by Breitbart and are listed in Table 2. As you might expect, those with far advanced disease are more likely to kill themselves than are those who are undergoing active treatment to cure the disease. Exceptions to this exist. An oncologist who had prac-

TABLE 2. CANCER SUICIDE VULNERABILITY FACTORS

1. Advanced illness; poor prognosis
2. Depression; hopelessness
3. Pain
4. Delirium
5. Loss of control; helplessness
6. Pre-existing psychopathology
7. Prior suicide history; family history
8. Exhaustion; fatigue

ticed for nine years told me that only two of his patients had ever killed themselves, and each had recently been diagnosed as suffering early-stage disease. One was a young man undergoing potentially curative chemotherapy for cancer of the testis.

Depression and hopelessness are the major predisposing conditions for suicide in people who don't have cancer. These factors also lead cancer patients to think that life is not worth living. An underlying depression may be harder to spot in cancer patients since symptoms of depression—such as poor appetite, insomnia, sadness, and withdrawal—may also be directly related to the cancer. Poor control of symptoms contributes greatly to depression. Fortunately, depression in cancer patients can be treated with a combination of good management of physical symptoms, psychosocial support and, at times, psychotherapy and antidepressive medications.

Among the cancer patients that I have seen who have actively wanted euthanasia or suicide, pain has been the dominant problem. My experience may be somewhat biased because my professional life is largely spent operating a cancer pain service and I see an unending stream of patients suffering pain that is difficult to control. However, throughout the literature of this topic, uncontrolled pain is the single factor most often cited as the reason for a cancer patient to commit suicide. As I will discuss in Chapter 5, cancer pain *can* be controlled in the vast majority of patients, often with relatively simple techniques and medications.

Delirium or mental confusion plays a significant role in many cancer suicides, especially impulsive suicides in hospital, according to Dr. William Breitbart, a psychiatrist at Memorial Sloan Kettering Cancer Hospital in New York. About 20% of their suicidal cancer patients exhibited an organic brain syndrome or delirium at the time of their psychiatric evaluations. The Washington and California state euthanasia initiatives provided that those who are not of sound mind may not sign the documents to authorize euthanasia or assisted suicide. A significant proportion of these people may improve from their confusional state with changes in medications, or treatment of the cancer, or its complications.

Loss of control or feelings of helplessness are mentioned frequently as factors in suicidal cancer patients. No one wants to be a burden on family and friends at the end of life. Loss of mobility, paraplegia (loss of leg function), loss of bowel and bladder function, amputation, inability to talk, loss of sensation, and inability to eat or swallow all tend to increase the risk of suicide in cancer patients. Psychological distress or the disturbances in interpersonal relationships arising from these problems may magnify them, and bring one to consider suicide.

Having a terminal illness does not mean that you will necessarily become crazy or need a psychiatrist to help you deal with your feelings. However, pre-existing psychiatric disorders do occur in a proportion of terminally ill patients and this surely raises the risks for suicide. Patients with schizophrenia, or previous episodes of severe depression and/or suicide attempts, will be at still greater risk for suicide when they develop cancer. Others at higher risk include people with personality disorders, and especially those suffering from alcoholism and substance abuse. These problems often call for further psychiatric treatment.

Fatigue and exhaustion are also factors that predispose cancer patients to increased risk for suicide. Patients now undergo ever more elaborate diagnostic studies, and then active anticancer treatment regimens that include surgery, radiotherapy, chemotherapy, and biological response modifiers. This can, and usually does, drain the physical, emotional, and financial energy of patient *and* family. If family and friends now withdraw under the strain, the patient's increased feeling of isolation and abandonment readily evolves into suicidal thinking and suicide.

Hospice team support is essential in helping patients and their loved ones cope with the ravages of cancer. If the family is unsupported by hospice nurses, social workers, and volunteers, caring for a loved one can prove overwhelmingly difficult. An experienced, skilled hospice team can help patients and families deal with cancer-caused emotional deficits before they escalate into reasons for suicide.

Dr. Breitbart notes that spouses, parents, and other family members of cancer patients may also be at increased risk of suicide because of the stress of the situation. Again, a hospice team is needed to spot potential problems as well as to provide support for the entire family.

CHAPTER 3

Patients Who Have Requested Euthanasia

WILLIAM SONG

William Song came to Los Angeles County–USC Medical Center with severe pain in his right shoulder, arm, and hand. Diagnostic evaluation determined the cause to be lung cancer. Surgery was not possible because the cancer had spread too far. He had lost over 20 pounds in the previous six months and was growing steadily weaker.

After numerous X-rays and other tests, his physicians prescribed radiation treatments and intravenous morphine for relief of his pain. Then they asked me to evaluate Mr. Song and make recommendations to improve his pain control.

Mr. Song was frightened of his disease, of the pain, and of dying. His fear and pain came not only from the cancer growing relentlessly in his body, but also from deep and complex emotional and psychological roots.

Ten years before his illness began, he had abandoned his family in Korea, emigrated to the United States, and was now living with cousins in Los

Angeles. His cousins were concerned about Mr. Song, but they had busy work schedules, so they had very limited time to take him to medical appointments and care for him. Mr. Song felt that he was a burden on his cousins and had no one else to depend on.

After talking to him and reviewing his records, I saw that the continuous intravenous morphine his doctors had prescribed was ineffective and impractical. It didn't control the pain, and it tied him to an intravenous catheter that had to be changed every few days. Intravenous medication was also unnecessary since he could take medications by mouth.

I prescribed an oral form of long-acting morphine to be given each 12 hours in gradually increasing doses. In addition, regular morphine tablets were administered whenever his pain worsened or he suffered episodes of "breakthrough pain." I also added anti-inflammatory and antidepressant drugs to his ongoing regimen as helper pain medicines, or co-analgesics.

Freed from his intravenous catheter and with tests completed and radiation underway, Mr. Song was soon discharged from the hospital and scheduled to complete radiation therapy as an outpatient. However, he still felt depressed, hopeless, and increasingly unsupported by his cousins. He began to cough up blood, which added to his fears. The radiation therapy treatments caused nausea and vomiting that was only partially controlled by anti-nausea medication.

After the second week of radiation treatment, he asked the pain service nurse for enough medication to kill himself. The nurse told him that we would not help him kill himself, but that we would continue to do our best to control his pain. She spent many hours with him, listening to him talk about his fear and his feelings of helplessness. She kept his cousins informed about his condition and tried to increase their involvement in his care.

I asked the consultant psychiatrist to our Cancer Pain Service to see Mr. Song. The psychiatrist provided therapy for him almost daily for several weeks, focusing on Mr. Song's guilt about abandoning his family, as well as his own feelings of being abandoned. As the counseling progressed, Mr. Song developed the courage to write one of his sons in Korea in an effort to reestablish their relationship.

Gradually, Mr. Song's physical and emotional condition began to improve. The radiation therapy reduced the size of his tumor and he stopped coughing up blood. The pain medicines—together with the radiation and psychotherapy we were providing—finally controlled his pain. His depression lifted, and he no longed wanted us to assist in his suicide.

He received a letter from his son inviting him to come back to Korea to spend the rest of his days. The ward social worker helped him make the arrangements. Mr. Song looked forward to the moment he would be reunited with his family. The realization that his time was limited seemed to make him enthusiastic about experiencing each day fully.

MR. MICHAELS

Mr. Michaels, a 45-year-old AIDS patient, dropped out of medical school after one year to work as a chemist and computer programmer. He found that he was infected with HIV when his common-law wife tested positive. In retrospect, the only possible times that he believed he might have been exposed to HIV were when he had dressed a very bloodly wound for an injured friend, and during a single homosexual experience.

He survived a bout of pneumocystis pneumonia. However, other infections struck in rapid succession over about nine months. These included mycobacterium avium intercellulare (an organism similar to tuberculosis), candida esophagitis (a fungus infecting the swallowing tube), cytomegalovirus retinitis (threatening blindness), and cryptosporidium (causing diarrhea).

The HIV attacked the nerves leading to his right arm, causing a severe pain that interfered with his sleeping and eating. When hospitalized for the placement of a catheter, his pain increased dramatically, requiring hydromorphone (an opioid analgesic). On discharge from the hospital, the hydromorphone was not continued, with no explanation. Mr. Michaels' physician-assistant in the AIDS clinic asked me to help with treatment of the pain. I represcribed the hydromorphone in conjunction with helper pain medicines (co-analgesics), which subsequently relieved his pain.

Mr. Michaels was scheduled to enroll in one of my research study protocols when he suddenly developed vomiting and diarrhea. The AIDS clinic doctors admitted him for hydration and diagnostic tests. His blood cultures showed that the catheter was infected. Antibiotics slowly cleared the infection, but he remained profoundly weak and wasted. He weighed barely over 100 pounds.

The nurse with whom I work saw him shortly after his hospital admission and reported that he was talking about euthanasia and suicide. One evening a few days later, I visited him on the infectious disease ward of the hospital. His common-law wife and a friend arrived later. After a few minutes, he asked his friend to take a walk so that he could discuss his situation with his wife and me.

Mr. Michaels detested the hospital. He complained about the long waits for the nurses to clean him after each bout of diarrhea. The wasting away of his body bothered him. I found that his doctors at my hospital did not recommend inserting a feeding tube into his upper intestine or feeding him by vein (total parenteral hyperalimentation). He asked me if he should request a transfer to another hospital where they would employ these high-tech feeding techniques. I pointed out the discomfort and potential complications of these feeding techniques and agreed with his doctors not to recommend them.

He then mentioned that he had heard of using a quinidine-type drug to overdose and cause an "eternal sleep." I told him that I could understand what he was feeling, since I was writing a book on euthanasia and assisted suicide. I pointed out that there was another alternative to consider, that he didn't have to choose between either all-out efforts to prolong his life, such as hyperalimentation, or hastening his death with a drug overdose.

I asked what he knew about hospice care. Mr. Michaels' wife volunteered that she had visited a hospice that day to see whether it was suitable for him, and had been very favorably impressed with the setting and the staff. A bed in the hospice would be available for Mr. Michaels in a few days. I spoke to them about the hospice approach, and its focus on the person's symptoms, both physical and emotional, during a terminal illness. I also mentioned the hospice emphasis on support for the caregivers, such as Mr. Michaels' wife. Hospice, I suggested, could give him and his loved ones more comfort for whatever time remained in his life.

He changed the subject and asked his wife to invite his father to visit. I asked Mr. Michaels about his relationship with his father. He responded that the man was emotionally cold and still had a very negative attitude about AIDS patients. The only time they could recall seeing his father cry was the previous year, when Mr. Michaels' mother had died of cancer. He wondered out loud whether a visit would be too stressful for his father, who himself was soon to undergo coronary artery bypass surgery. He questioned whether his father would come at all, but his wife reassured him that he would indeed come, and that the visit would be good for both of them.

Mr. Michaels asked me whether I would reorder the hydromorphone that had been discontinued during his hospitalization. His arm still hurt, although not as severely as before. He thought that the hydromorphone would help him relax and be more comfortable. I ordered the medication.

I next spoke with Mr. Michaels' wife about two weeks later, on the day of his death. She said that our conversation had assuaged much of his anxiety and that he had entered the AIDS hospice because she was unable to care for him at home without losing her job. But she had visited him almost every evening. I learned that during a phone conversation with his father, Mr. Michaels had finally told him that he had AIDS. For the first time in years, he also told his father that he loved him. This conversation seemed to give him peace and comfort.

By all accounts, pain was no longer a problem for Mr. Michaels. He no longer asked for a gastric feeding tube or total parenteral nutrition. He did not mention suicide or euthanasia again. In his final weeks, Mr. Michaels came to an acceptance of his situation. He shared many of his feelings with his wife, who greatly appreciated their last days together.

Mr. Michaels and his loved ones would have missed some powerfully healing experiences if he had committed suicide, as he had once considered.

MR. WHITE

Mr. White was a 62-year-old white man with lung cancer diagnosed six months before he was referred to our Cancer Pain Service. His oncologists had treated him with radiation therapy to the right upper lung three months previously. He opted for no additional antitumor therapy.

Mr. White's doctors insisted that he come to the clinic before they would renew his morphine prescription. Mr. White was quite weak and required a wheelchair for traveling even moderate distances. His wife had a weak leg, which made it hard for her to push his wheelchair. Attending oncology clinic was a major ordeal for both of them. They didn't understand the necessity of further oncology clinic visits anyway since he had declined to have chemotherapy. A nurse in the oncology clinic sought to end this stalemate, so she asked whether I would reorder his pain medication.

I called Mr. White at home to obtain the details of his medical history and see whether I might help him. After listening to his story, I renewed his morphine prescription at a higher dose. We also agreed upon a convenient time for me to visit him at home to finish my assessment.

When I arrived at his home, he said that his pain was slightly improved. For about 12 months before his referral to the Cancer Pain Service, he had experienced severe pain in the right shoulder, right arm, and forearm. Radiation therapy had greatly decreased the pain for several weeks; however, with tumor regrowth, the pain had again become intense despite substantial doses of oral morphine.

Because of his pain and the prognosis of the tumor, he stated that he had nothing to live for and was considering suicide. I could see why he felt the way he did. His pain was not well controlled, and he did not feel supported by his doctors at the hospital.

I questioned Mr. White further about his suicidal thoughts. He said that he knew that there was no cure. Since the disease was so painful and such a burden on his wife and family, he wondered why he shouldn't get it over with instead of letting it drag out. I asked whether he had formulated specific suicide plans. He said that he had no plans, but was thinking about doing it after Christmas. Mr. White's wife, who was also in the room, showed obvious apprehension in discussing this topic, and we moved to other issues.

After examining Mr. White, I detailed the treatment of his pain, which he only partly understood. I again increased the morphine dose and also added prednisone as a corticosteroid helper pain medication. Corticosteroids like prednisone often boost appetite and overall sense of well-being, as well as helping pain. I also added nortriptyline, another helper pain drug that additionally served to treat his depression and insomnia.

I explained the hospice philosophy of providing symptom control and emotional and psychological support for the patient and loved ones. I assured him that further hospitalizations or clinic visits would not be necessary because we could provide all the needed support at home.

I told him I would order a hospice nurse to visit him at home on a regular basis. The hospice nurse, I told them, would monitor his pain and other symptoms, check his medications, make sure they were working, report to me about any changes needed, arrange for a hospital bed and other needed appliances to be brought to the home, answer questions, and generally support them throughout the process of his illness. Mr. White and his wife expressed great appreciation to me for introducing them to the hospice program and explaining everything to them.

Over the next weeks, Mr. White's pain was much better controlled. Because of the wife's busy schedule, it took about a week for the hospice nurse to visit. However, the nurse provided them considerable support. The Whites enjoyed Christmas with their large extended family (about 60 people).

Mr. White died peacefully a week later of pneumonia. During his last few weeks, he had made no further mention of suicide. His wife had done a fine job caring for him at home; without her help, he would have had to go into a hospital or nursing home, which he dreaded. Ms. White told me that it gave her comfort that she cared for him at home despite how difficult it had proved to be. She expressed that his final weeks were very meaningful to him and to all his family.

"It's Over Debbie"

From a Physician's Letter Admitting Euthanasia

The call came in the middle of the night. As a gynecology resident rotating through a large, private hospital, I had come to detest telephone calls, because invariably I would be up for several hours and would not feel good the next day. However, duty called, so I answered the phone. A nurse informed me that a patient was having difficulty getting rest, could I please see her. She was on 3 North. That was the gynecologic–oncology unit, not my usual duty station. As I trudged along, bumping sleepily against walls and corners and not believing I was up again, I tried to imagine what I might find at the end of my walk. Maybe an elderly woman with an anxiety reaction, or perhaps something particularly horrible.

I grabbed the chart from the nurses' station on my way to the patient's room, and the nurse gave me some hurried details: a 20-year-old girl named Debbie was dying of ovarian cancer. She was having unrelenting vomiting apparently as the result of an alcohol drip administered for sedation. Hmmm, I thought. Very sad. As I approached the room I could hear loud, labored breathing. I entered and saw an emaciated, dark-haired woman who appeared much older than

20. She was receiving nasal oxygen, had an IV, and was sitting in bed suffering from what was obviously severe air hunger. The chart noted her weight at 80 pounds. A second woman, also dark-haired but of middle age, stood at her right, holding her hand. Both looked up as I entered. The room seemed filled with the patient's desperate effort to survive. Her eyes were hollow, and she had suprasternal and intercostal retractions with her rapid inspirations. She had not eaten or slept in two days. She had not responded to chemotherapy and was being given supportive care only. It was a gallows scene, a cruel mockery of her youth and unfulfilled potential. Her only words to me were, "Let's get this over with."

I retreated with my thoughts to the nurses station. The patient was tired and needed rest. I could not give her health but I could give her rest. I asked the nurse to draw 20 mg of morphine sulfate into a syringe. Enough, I thought, to do the job. I took the syringe into the room and told the two women I was going to give Debbie something that would let her rest and to say goodbye. Debbie looked at the syringe, then laid her head on the pillow with her eyes open, watching what was left of the world. I injected the morphine intravenously and watched to see if my calculations of its effects would be correct. Within seconds her breathing slowed to a normal rate, her eyes closed, and her features softened as she seemed restful at last. The older woman stroked her hair of the now sleeping patient. I waited for the inevitable next effect of depressing the respiratory drive. With clock-like certainty,

within four minutes the breathing rate slowed
even more, then became irregular, then ceased.
The dark-haired woman stood erect and seemed
relieved.

It's over, Debbie.

Name Withheld by Request

In January 1988 the above letter appeared in
the *Journal of the American Medical Association*
(JAMA).[1]

Members of the editorial board of *JAMA* were
divided about the ethics of printing this anonymous
letter. They also worried about the possible legal
implications of publishing an admission of a murder without revealing the source. Finally, Dr. George
Lundberg, former California pathologist and present
chief editor of the *JAMA,* made the unilateral decision to print the letter. He felt the issue of euthanasia for terminally ill people should be widely debated
and resolved. He said that euthanasia in this kind
of situation could occur, "in almost any hospital, in
almost any community, in the United States."

The county prosecutor in Chicago, the location
of corporate offices of the American Medical Association and the *JAMA,* subpoenaed Dr. Lundberg
to supply further details about the author of the
letter describing euthanasia. Dr. Lundberg and the
AMA refused, citing the Illinois Reporter's Privi-
lege Act which protects journalists from being forced
to reveal their sources. Dr. Lundberg and the *JAMA*
prevailed in court.

The credibility of this account has been widely questioned. However, if this is a true story, it underscores the lack of training and knowledge in the area of cancer pain management.

Any person with chronic cancer-related pain should have been given regular doses of morphine or other strong opioid. If this had been done before the anonymous physician saw the patient, then 20 milligrams of IV morphine would not have been a fatal overdose. Cancer patients who regularly take morphine or other opioids rapidly develop a tolerance to higher doses that would kill people who had not been previously taking opioids. Twenty milligrams of IV morphine would be possibly lethal only in a cancer patient who had previously taken little or no opioid. I have cared for patients who required over 100 milligrams per hour and showed no signs of respiratory depression.

If Debbie was killed with only 20 milligrams of IV morphine, then she previously suffered with untreated pain from the cancer. This suggests that her request for euthanasia was because of inadequate pain treatment. Her pain, like the pain of most cancer patients, could probably have been controlled with adequate medication.

When this issue surfaced in the spring of 1988, newspaper articles focused on the legal implications of the case, speculation about how common the practice of euthanasia is now, the chances of a physician who practiced euthanasia being discovered and prosecuted, opinions and practices about euthanasia in other countries, and opinion polls here on the subject. Unfortunately, the possibility that Debbie's pain had been inadequately treated, owing to the physician's lack of adequate training in pain management, did not receive much media attention.

By publishing "It's Over Debbie," Dr. Lundberg may have hoped to motivate doctors to openly discuss whether euthanasia should be legalized, and whether they are perhaps already performing "mercy killings." He obviously favored a change to permit euthanasia in some circumstances.

Dr. Sigmund Freud

Biographer, Ernest Jones, described Dr. Sigmund Freud's illness with cancer of the oral cavity, first discovered in 1923. In 1928, Dr. Max Schur became Freud's personal physician. He treated him both in Vienna and London until Freud died in 1939 at the age of 82.

When he first engaged Schur as his physician, Freud extracted the promise of euthanasia[1]:

"...Mentioning only in a rather general way 'some unfortunate experiences with your predecessors,' he expressed the expectation that he would always be told the truth and nothing but the truth. My response must have reassured him that I meant to keep such a promise. He then added, looking searchingly at me: 'Promise me one more thing: that when the time comes, you won't let me suffer unnecessarily.' All this was said with the utmost simplicity, without a trace of pathos, but also with complete determination. We then shook hands.

"Thus doctor and patient were under euthanasia[2] commitment during approximately the last 11 years of Freud's life."

After 16 years of struggling with cancer, and 33 operations, Freud lay on his deathbed in London at the age of 82. When he felt the futility of prolonging life further, he asked his personal physician to ease his way. Biographer Jones related Dr. Schur's description of the final scene:

"On the following day, September 21, 1939 while I was sitting at his bedside, Freud took my hand and said to me, 'My dear Schur, you certainly remember our first talk. You promised me then not to forsake me when my time comes. Now it's nothing but torture and makes no sense any more.'"

"The next morning Schur gave Freud a third of a grain of morphia (20 milligrams of morphine). For someone at such a point of exhaustion as Freud then was, and so complete a stranger to opiates, that small dose sufficed."

Freud's situation was quite similar to "Debbie's" in the preceding case, down to the same relatively small dose of morphine. Previous to his euthanasia, Freud's pain had not been appropriately treated with opioid analgesics. If he had received reasonable doses of morphine to control chronic pain, he probably would not have asked for "aid in dying."

CHAPTER 4

The Right to Die

Choice in Dying (formerly Concern for Dying and the Society for the Right to Die) beautifully expresses the essence of the right-to-die movement in its brochure: "The Society ...

- Believes that the basic rights of self-determination and of privacy include the right to control decisions relating to one's own medical care.
- Opposes the use of medical procedures which serve to prolong the dying process needlessly, thereby causing unnecessary pain and suffering and loss of dignity. At the same time we support the use of medications and medical procedures which will provide comfort care to the dying.
- Recognizes that a terminal condition may prevent a patient's participation in medical care decisions and that previously expressed wishes may not be honored by physician and/or hospital.
- Sees to: (1) protect the rights of a dying patient, and (2) protect physicians, hospitals and health care providers from the threat of liability for complying with the mandated desires of those who wish to die with medical intervention limited to the provision of comfort care.[1]

Choice in Dying does not advocate euthanasia or assisted suicide as a means to achieve death with dignity.

ORDINARY VERSUS EXTRAORDINARY MEDICAL TREATMENT

In all medical training, the physician is taught to act in the best interests of the patient. This means the provision of different treatments to generally healthy patients from those given to terminally ill patients. Most of the terminally ill prefer to die at home surrounded by friends and family rather than in the modern high-tech hospital. This usually requires treating the patient's symptoms with "low-tech" therapies. The distinction between ordinary and extraordinary treatment often depends on the patient, the doctor, and any of many other variables. It is of particular importance for comatose or other incompetent patients who cannot decide such issues themselves. Physicians consider the following factors when determining what is ordinary and what is extraordinary treatment:

- usual vs unusual treatment for a given condition
- simple vs high-tech treatment
- invasive (such as surgical) vs noninvasive treatments (such as medication)
- inexpensive vs expensive treatment
- conservative vs high-risk or long-shot therapies

The courts have not yet reached a consensus on just what the critical distinctions may be or how to evaluate particular cases in light of these distinctions. For example, antibiotics may ordinarily be used for pneumonias or other life-threatening infections. However, such antibiotics, which are intended to prolong life, may or may not improve the quality of life of terminally ill cancer patients. Thus, whether the use of these antibiotics is ordinary or extraordinary should be individually determined, based on a discussion of the situation between the primary care physician and the patient or patient's next of kin. Certainly respirators and dialysis machines constitute extraordinary measures of life support and, for terminally ill cancer and AIDS patients, would add to the suffering and needlessly prolong the dying process.

FORCE FEEDING AND HYDRATING THE TERMINALLY ILL

Many hospice workers have noticed over the years that, in actively dying or comatose people, forced feeding and hydration does not relieve a patient's distressing symptoms. Indeed, it often increases the distress from the pain and adds the discomfort of the tubes. The artificially administered fluids accumulate in the upper respiratory tract, producing increased secretions, and in the lower respiratory tract, producing cough and shortness of breath. The fluids also may accumulate in the abdomen and legs, causing additional discomfort.

If the patient has far advanced disease and appears to be close to death, I usually recommend not placing feeding tubes and IV lines or taking

* * * * *

*...polls show that most people, if they
were terminally ill and comatose,
would want to have tubes for feeding
and hydration discontinued.*

* * * * *

them out if they are in place. Results of polls show that most people, if they were terminally ill and comatose, would want to have tubes for feeding and hydration discontinued.

Despite the polls, physicians, lawyers, philosophers, and theologians have widely debated whether comatose, terminally ill people should be force-fed and given intravenous hydration. After some deliberation, the Los Angeles County Medical Association, together with the LA Bar Association, published a white paper on this issue in 1987. It concluded that, in certain cases, there is no legal or moral imperative to provide terminally ill patients with tube feeding or IV hydration.[2]

The legal requirement of forced feeding and hydrating terminally ill, comatose people was not addressed in the Washington state Natural Death Act of 1979. The 1991 "Death with Dignity" initiative in Washington state (Initiative 119) specified which life-sustaining procedures may be withdrawn, and included both forced feeding and hydration. If this issue had not been linked to legalizing euthanasia, I, for one, would have voted for it.

Do Not Resuscitate

Cancer patients with curable diseases, or those with good prospects for prolonged remission, should, if needed, be given cardiopulmonary resuscitation (CPR) and treated with life-support technology. If cardiac arrest occurs in such a cancer patient, either because of trauma or after potentially curative surgery, heart attack, or other problem unrelated to the cancer, high-technology intensive care treatment is appropriate, and the patient should be resuscitated. Patients with advanced disease who enjoy little chance of achieving a good quality remission generally should not be given CPR or life-support technology.

I have seen some very sick cancer patients with potentially curable diseases who achieved good remissions because of high-technology, intensive care treatment. However, I have rarely, if ever, seen a patient with advanced or relapsed cancer who benefited from CPR and life-support machines. The heart may well restart, only to fail again after the patient has spent several agonizing days on a respirator in an ICU.

In the vast majority of terminally ill cancer or AIDS patients, there is no reason to attempt cardiopulmonary resuscitation when the heart stops. This has no medical purpose except as a last desperate act of denying death—and it wastes expensive health care resources.

* * * * *

*In the vast majority of terminally ill
cancer or AIDS patients...CPR...has no
medical purpose except as a last desperate
act of denying death—and it wastes
expensive health care resources.*

* * * * *

Most people suffering incurable cancers would rather die at home, surrounded by friends and family, than in a hospital. If they must die in the hospital, they would prefer to be placed in a quiet room rather than in the intensive care ward. Even if for some reason they had to be treated in an intensive care ward, they would prefer to go peacefully, rather than in a frenzy of electrical shocks and chest compressions.

If a terminally ill patient in an emergency situation is placed on a respirator, the physician may stop the respirator at the patient's or proxy's request. "Pulling the plug" in this situation is perfectly legal and appropriate—not at all an act of active euthanasia. If a respirator or another piece of advanced life-support equipment is stopped, the patient is not being killed, but simply being allowed to die naturally of the disease.

The decision to discontinue life-support technology usually comes only after it is apparent that further heroic treatment would prolong the patient's suffering and dying, without offering a realistic chance at remission or recovery. No one would want to remain on a respirator indefinitely once advanced cancer had caused their organs to fail. Courts have ruled that there is no legal or moral distinction between foregoing life-sustaining treatment and stopping treatment that has been started.

PHYSICIAN–PATIENT/FAMILY DISCUSSION OF THE CPR ISSUE

Since recent court cases have dealt with both foregoing and withholding life-support technology, hospitals have been forced to develop policies concerning the withholding of CPR and life-support machines. Before physicians can write orders to withdraw life-support technology, as well as not to resuscitate patients, they must obtain informed consent from the patients or next of kin.

Unfortunately, physicians dread having to discuss dying with cancer patients and their families. Such conversations almost seem to be a negation of the physician's role as healer and tantamount to admitting that their patients' situations are hopeless. Many times doctors simply neglect to discuss CPR with their terminally ill patients and their loved ones. The consequences of this omission are often disastrous. If physicians were properly trained in palliative care, this communication problem would be greatly diminished.

There are good and bad ways to approach the delicate topic of CPR. Given a terminally ill individual, the physician may recognize that prolonging the dying process on life-support machines would be cruel and pointless. Even so, the physician may not be able to convey this concern in a compassionate manner.

For instance, if the husband is unconscious, the doctor may ask the wife whether or not she would like them to perform CPR and use life-support machines. In this situation, the doctors interpret their obligation to discuss CPR and life-support technology as meaning that the decision is left solely in the hands of the family.

Trying to persuade the wife to decline CPR for her husband, the physician may begin a discussion of CPR by saying, "We have no further medical treatments that can save your husband. We have nothing more to offer. Do you want us to perform CPR and use breathing machines if his heart stops beating?" This approach places the already emotionally exhausted wife in a terrible dilemma. If she requests no resuscitation, she may always wonder whether her husband might have been saved with more effort. If she insists on resuscitation and her husband lives an extra week on an ICU respirator, she may feel guilty for needlessly prolonging his pain and suffering.

Because of fear, anxiety, guilt, or sheer panic, the wife or other family members may insist on CPR and the use of all available medical technol-

* * * * *

*...I recommend...to first begin discussing
the question of resuscitation with cancer
patients and their families as soon as
it appears that anticancer treatments
are not controlling the disease.*

* * * * *

ogy. To them this gesture says "no" to hopelessness
and "no" to abandonment. Sometimes, out of a sense
of guilt, relatives may demand that no medical technology be spared to keep the patient alive. Unfortunately, this decision creates additional misery for
all concerned.

A better approach, and one that I recommend,
is to first begin discussing the question of resuscitation with cancer patients and their families as
soon as it appears that anticancer treatments are
not controlling the disease. Patients can even be
healthy, relatively speaking, and not at all close to
death, when resuscitation is first discussed. In fact,
people are relieved to discuss this topic, knowing
that they retain at least some control over whether
they will die naturally.

Here is how I work toward decisions on the
best care for my patients. First, I consider the patient's overall prognosis, quality of life, and chances
of leaving the hospital alive. Then I decide whether
I think resuscitation would be appropriate if the
patient's heart should stop beating.

Next, I bring the question up in a patient care conference to elicit the feelings and opinions of all members of the medical team, including doctors, nurses, social workers, and clergy. After reaching a consensus, I arrange a meeting with the patient and family.

In any discussion of CPR, the physician must be sensitive to the feelings of both the patient and the loved ones. Before beginning to discuss CPR with them, it is essential that they know the status of the disease and the prognosis. Sometimes, patients may not be aware of the seriousness of their situation. It may even take several days to inform them adequately about the disease process since they may be able to comprehend and assimilate only a little bit at a time.

When CPR is not deemed appropriate by the medical staff, I begin informing the patient and family by emphasizing what the medical staff is planning to recommend for the patient, rather than what they do not recommend. I usually say, "Though we don't have a treatment that can cure the cancer, we have good treatments to help control your pain and to handle any other troublesome complication of the disease or its treatment."

I dispel their fear of abandonment by saying "We will visit you frequently and will be available any time you need help." I go on to say that the medical staff has fully discussed the situation of a possible cardiac arrest. I tell them that the staff doesn't believe that heroics, such as the use of CPR

* * * * *

By entering the hospice program, the patient and family have chosen symptom control and psychosocial support over medical heroics.

* * * * *

and high-tech life-support technology, would be of benefit.

I always inform them about CPR, but without asking them to say "yes" or "no." In earlier days, physicians alone had always decided what course of action to recommend. In this fearful, vulnerable time, the patient and family do not need the additional burden of deciding about CPR in the absence of any thoughtful guidance and recommendations from a trusted physician.

The patient or family may of course disagree with the medical staff's recommendation. However, when they are fully informed about the disease and the approach outlined above is used, I have rarely seen people go on to insist on CPR. Instead of being more fearful and anxious, the patient and family usually appear comforted and relieved after these discussions.

If the patient is already in a hospice program—which I firmly believe most advanced cancer patients should be—the question of CPR has already been determined. By entering the hospice program, the patient and family have chosen symptom control and psychosocial support over medical heroics. Ideally, all physicians treating cancer patients should be trained in the hospice approach. Needless resuscitations and ICU admissions would then be prevented, as well as unnecessary suffering of dying patients and their families and loved ones.

RIGHT TO DIE— COURT DECISIONS AND LEGISLATION

Between 1920 and 1985, fifty-six "mercy killing" cases were brought into the court for trial. Of these, the courts in the United States found only ten defendants guilty and imprisoned them. Twenty defendants were found guilty and given suspended sentences or probation. Fifteen defendants were acquitted, and the court dismissed six cases.

Four of the cases involved children, which meant to the court that they were victims of involuntary euthanasia. I will briefly review two of the most important court cases.

KAREN ANN QUINLAN

In 1976, at the age of 22, Karen Ann Quinlan overdosed on drugs and was rushed to a local New Jersey hospital, and placed on life-support machines. Unfortunately, her brain had been without oxygen for so long that she did not wake up. Although she was brain-dead, it was necessary for a court to rule that she could be taken off the respirator before her physicians could do so. After years of legal battles, the court allowed doctors to discontinue the respirator. Ms. Quinlan continued to breathe, however, and lived in a nursing home in a chronic vegetative state for several more years before eventually succumbing.

Fortunately, the Quinlan case precedent has made it possible to withdraw life-support equipment from people determined to have suffered brain death and without apparent hope of recovery. In the vast majority of such cases, we no longer need to go through the court system.

The Washington state "Death with Dignity" initiative addressed part of Ms. Quinlan's situation in that a chronic vegetative state would have been defined as terminal. If the initiative had passed, someone in Karen's situation who had a living will requesting the withdrawal of forced feeding and hydration in such a situation could have expected the request to be honored. This kind of request from a nonbrain-dead person, even one with a living will petitioning family members, still may or may not be honored. Further legal clarification is needed in Washington state and most other states.

The Quinlan case would not have been affected by initiatives or legislation to legalize euthanasia or physician-assisted suicide since no one ever proposed to kill Ms. Quinlan with an intentional overdose of drugs.

THE WILLIAM BARTLING CASE—PHYSICIAN REFUSAL TO STOP LIFE SUPPORT

Mr. Bartling was hospitalized in the Glendale Adventist Hospital in 1984 with severe heart and lung disease. During his hospitalization, a lung biopsy led to the diagnosis of inoperable lung cancer and put him into respiratory failure by collapsing his lung. He required a mechanical respirator.

When Mr. Bartling understood his diagnosis and prognosis, he asked that the respirator be turned off and he be allowed to die. His physician, backed by the hospital administration and legal counsel, did not comply with his request. The case went to court and the judge decided for the physician and hospital. After Mr. Bartling's death, the appellate court correctly decided that he had every right to have the life-support machines turned off.

Existing law already covered Mr. Bartling's situation as a terminally ill person asking for withdrawal of life support. Unfortunately, his doctors did not make this determination themselves, and insisted on involving the court. The doctors could have taken the issue to the ethics committee of the hospital, which should have honored Mr. Bartling's wish to stop life-support treatment.

Many people are confused on this point about a patient's right to decline life support and think that new laws, such as Washington's now defeated Initiative 119, are needed to allow physicians to turn off respirators for terminally ill patients who request it. (*See also* the Appendix for Ms. Bartling's solicitation letter in behalf of the Americans Against Human Suffering.)

Given the facts of the case, I don't know why the lower court held that he had to remain on the respirator against his will. The appellate court made the right decision according to existing law. In this tragic case, the physicians, the hospital administrators, the lawyers, and the lower court each share the blame, but not the law.

THE NANCY CRUZAN CASE

Nancy Cruzan was a 32-year-old Missouri resident who existed in a "vegetative state" for seven years due to brain damage from a car accident. She had no higher brain function and required total nursing care, including a feeding tube and catheter draining her bladder. Her parents requested that the feeding tube be removed and that she be allowed to die. One of Nancy's friends testified that Nancy told her several times that she wouldn't want to live as a vegetable.

The Missouri Supreme Court ruled 4-3 that the evidence of Nancy's wishes was unreliable and denied the family's request for discontinuation of the feeding tube. The Court declared that, since Nancy Cruzan did not have a "living will" or other documentation of her wishes if she were in a chronic vegetative state, no one could say what she would want. By a 5 to 4 margin, the US Supreme Court declined to intervene in the individual states' rulings on this issue.

Nancy's parents eventually took her to a nursing home in another state. Physicians and nurses in that state were permitted by the courts to comply with the family's wish to have forced feeding and water stopped. She died within two weeks.

With the Cruzan case, the Court declined the opportunity to extend the benefits of existing law and medical practice for the terminally ill to those in an irreversible persistent vegetative state. I was disappointed with this decision. However, legalizing active euthanasia would not be an appropriate response to this ruling. There is a big difference between allowing a person to die by not starting or withdrawing medical support ("passive euthanasia") and actively killing someone by injection of an overdose of medication ("active euthanasia").

If you feel that, if you were in Nancy Cruzan's condition, you would like to be allowed to die, telling your family and close friends does not guarantee that your wishes will be honored. You must fill out "living will" and "durable power of attorney for health care" documents explicitly stating your wishes in this situation.

CHAPTER 5

Traditional Arguments Against Euthanasia

In my view the main argument against legalizing active euthanasia is that hospice care is a far preferable alternative. In the following I will present a synopsis of the classical arguments against euthanasia for completeness, and because some offer valid related points.

THE "SLIPPERY SLOPE" ARGUMENT

Kamisar[1] first formulated the "slippery slope" argument in 1958. This grants that there may be a few cases in which active euthanasia is appropriate, but that legalization of euthanasia in any form would cause far-reaching, harmful social and legal consequences. Along with the justifiable cases of terminally ill people asking for and receiving a quick, merciful death, there would inevitably be cases in which euthanasia would be clearly wrong. Indeed, euthanasia could be seen as the first step toward adopting Nazi-style policies of killing the old, weak, and socially disfavored. Since this path is slippery and steep, we should stay off it altogether.

ABUSE AND THE LEGALIZATION OF EUTHANASIA

A law legalizing euthanasia might well be abused, with some person's life being ended against his or her consent for a motive other than mercy.

I see no reason for a physician using a legalizing law to hasten the death of a terminally ill person who does not want euthanasia. Doctors generally strive to preserve life with all their skills. They are very uncomfortable with death. Doctors' fees depend on treating live patients, so there is no economic reason to abuse this proposed law.

DIAGNOSES AND PROGNOSES MAY BE WRONG

Currently, for a cancer patient to elect the Medicare hospice benefit, two physicians must certify a prognosis of less than six months. A significant proportion of the patients who elect the Medicare hospice benefit outlive that prediction.

Predicting how long someone may live with cancer is very difficult at best. I never tell patients they have only a certain number of months to live. Some whom I expected to die in a few months have lived for years. And some have lived a much shorter time than we, the caregivers, anticipated. I am now caring for two patients who have had metastatic cancer for five years and one with metastatic cancer for ten years.

THE RIGHT TO DIE
WILL BECOME A DUTY TO DIE

Frail, disabled elderly people who are financial and emotional burdens on their families may feel some pressure to ask for euthanasia to spare their families further suffering. Governor Lamm of Colorado has publicly said that for economic reasons we must not only begin to consider rationing health care services, but even to encourage euthanasia for elderly, incapacitated people in nursing homes.

Currently, few terminally ill cancer patients either ask for euthanasia or commit suicide. This may be true in part because of a lack of social acceptance, as well as of advertising, for these solutions. If euthanasia were available for the asking, some relatives or heirs of terminally ill patients might begin to recommend it. Legalization would give euthanasia increased exposure and legitimacy, as well as more advertising among a population of extremely vulnerable people—patients and their relatives.

THE PATIENT/PHYSICIAN RELATIONSHIP WILL BE WEAKENED

Apart from moral or ethical considerations, I don't know how any physician could risk performing euthanasia because of the potential damage to his or her future relationships with patients. I could imagine seeing a newly diagnosed cancer patient and being asked if I "aid terminally ill patients in dying." The following are some potential questions that a patient might ask:

- How many times have you needed to perform euthanasia?
- What were the circumstances?
- Would you have any problem performing euthanasia on me if I requested it and the circumstances

satisfied the conditions of the law?
- What drugs would you use?
- What would the lethal drugs feel like before they caused death?
- How do relatives generally cope with the loss compared with patients who are allowed to die of the disease without "assistance?"

I believe that many patients would feel very ambivalent about being cared for by a doctor who performed euthanasia.

THE HIPPOCRATIC OATH

Physician "aid-in-dying" would also be a violation of the Hippocratic oath, which obligates the physician to both try to benefit his patients and to do no harm. Although this oath is primarily of historical interest and no longer administered to most medical graduates, it remains another dissuading factor in the physician's mind.

POSITIONS OF RELIGIOUS DENOMINATIONS

The positions of the major religions on voluntary euthanasia for the terminally ill were recently set forth in detail by a publication of The Park Ridge Center for the Study of Health, Faith, and Ethics.[2] All the denominations surveyed encouraged the practice of allowing terminally ill patients to die by withholding or withdrawing medical treatment when the burdens of treatment outweigh the benefits.

Roman Catholic Views

Roman Catholic theologians have reflected on matters of death and dying for centuries. The most recent statement of the official church came from the Sacred Congregation for the Doctrine of the Faith's 1980 Declaration on Euthanasia, approved by Pope John Paul II, which declares:

> Human life is the basis of all goods, and is the necessary source and condition of every human activity and of all society. Most people regard life as something sacred and hold that no one may dispose of it at will, but believers see in life something greater, namely a gift of God's love, which they are called upon to preserve and make fruitful. And it is this latter consideration that gives rise to the following consequences:
>
> 1. None can make an attempt on the life of an innocent person without opposing God's love for that person, without violating a fundamental right, and therefore without committing a crime of the utmost gravity.
> 2. Everyone has the duty to lead his or her life in accordance with God's plan. That life is entrusted to the individual as a good that must bear fruit already here on earth, but that finds its full perfection only in eternal life.
> 3. Intentionally causing one's own death, or suicide, is therefore equally as wrong as murder: such an action on the part of a person is to be considered as a rejection of God's sovereignty and loving plan. Furthermore, suicide is also often a refusal of love for self, the denial of the natural instinct

to live, a flight from the duties of justice and charity owed to one's neighbor, to various communities, or to the whole of society—although, as is generally recognized, at times there are psychological factors present that can diminish responsibility or even completely remove it.

Specifically about euthanasia, the document states:

By euthanasia is understood an action or an omission which of itself or by intention causes death, in order that all suffering may in this way be eliminated. Euthanasia's terms of reference, therefore, are to be found in the intention of the will and in the methods used.

It is necessary to state firmly once more that nothing and no one can in any way permit the killing of an innocent human being, whether a fetus or an embryo, an infant or an adult, an old person, or one suffering from an incurable disease, or a person who is dying. Furthermore, no one is permitted to ask for this act of killing, either for himself or herself or for another person entrusted to his or her care, nor can he or she consent to it, either explicitly or implicitly. Nor can any authority legitimately recommend or permit such an action. For it is a question of the violation of the divine law, an offense against the dignity of the human person, a crime against life, and an attack on humanity.

Protestant Positions

Some of the Protestant denominations that oppose voluntary euthanasia are the Lutherans, Mennonites, Methodists, Presbyterians, Mormons,

Jehovah's Witnesses, Episcopalians, Christian Scientists, Baptists, and Disciples of Christ. The rationale for rejecting the option of euthanasia or assisted suicide is based generally on the doctrine, "only God can give life and only God should take it away."

The Adventists have no official position on euthanasia. The United Church of Christ is debating whether or not to support legalization of euthanasia in some extreme cases.

Unitarian Endorsement of Euthanasia

The Unitarian Universalist Association (the combined Unitarian and Universalist churches) supports voluntary euthanasia. The 1988 Unitarian Universalist General Assembly issued the following statements in its document on "The Right to Die with Dignity":

> Guided by our belief as Unitarian Universalists that human life has inherent dignity, which may be compromised when life is extended beyond the will or ability of a person to sustain that dignity; and believing that it is every person's inviolable right to determine in advance the course of action to be taken in the event that there is no reasonable expectation of recovery from extreme physical or mental disability; and...
>
> *Whereas*, prolongation (of life) may cause unnecessary suffering and/or loss of dignity while providing little or nothing of benefit to the individual, and...
>
> *Whereas*, differences exist among people over religious, moral, and legal implications of admin-

istering aid in dying when an individual of sound mind has voluntarily asked for such aid; and...

Whereas, obstacles exist within our society against providing support for an individual's declared wish to die; and...

Whereas, many counselors, clergy, and health-care personnel value prolongation of life regardless of the quality of life or will to live;

Therefore Be It Resolved: That the Unitarian Universalist Association calls upon its congregations and individual Unitarian Universalists to examine attitudes and practices in our society relative to the ending of life, as well as those in other countries and cultures; and...

Be It Further Resolved: That Unitarian Universalists reaffirm their support for the Living Will, as declared in a 1978 resolution of the general Assembly, declare support for the Durable Power of Attorney for Health Care, and seek assurance that both instruments will be honored; and...

Be It Further Resolved: That Unitarian Universalists advocate the right to self-determination in dying, and the release from civil or criminal penalties of those who, under proper safeguards, act to honor the right of terminally ill patients to select the time of their own deaths; and...

Be It Further Resolved: That Unitarian Universalists advocate safeguards against abuses by those who would hasten death contrary to an individual's desires; and...

Be It Further Resolved: That Unitarian Universalists, acting through their congregations, memorial societies, and appropriate organizations, inform and petition legislators to support laws that will create legal protection for the right to die with dignity, according to one's own choice.[3]

Islam

Islam, which claims six million followers in the United States, opposes euthanasia. The Qur'an and Sunna, the authoritative texts of Islamic doctrine, do not speak specifically about euthanasia. However, according to Islamic law, God is the creator of life. Consequently, persons do not own their lives and have no right to end them or to ask others to do so.

Judaism

The four branches of Judaism—Orthodox, Conservative, Reform, and Reconstructionsist—all forbid active euthanasia. The ancient Torah and Talmud did not address euthanasia or assisted suicide. However, in recent years, rabbis have answered questions about death and dying in "responsas" that have come to be considered authoritative. For example, a responsa from the Reform Jewish tradition addressing euthanasia is the following:

> Human life is more than a biological phenomenon; it is the gracious gift of God, it is the in-breathing of His spirit. Man is more than a minute particle of the great mass known as society: his is the child of God, created in His image. "The spirit of God hath made me," avers Job in the midst of his suffering, "and the breath of the Almighty gives me life" (Job 33:4). Thus, human life, coming from God, is sacred, and must be nurtured with great care. And man, bearing the

divine image, is endowed with unique and hidden worth and must be treated with reverence.[4]

Greek and Russian Orthodox

The Greek Orthodox Church has no ancient doctrine addressing euthanasia. The modern Greek Orthodox ethicist, Stanley Harakas, speaks for the church saying, "The Orthodox Church completely and unalterably opposes euthanasia. It is a fearful and dangerous 'playing at God' by fallible human beings."

Likewise the Russian Orthodox Church rejects euthanasia as against God's will. Some years ago, a televised discussion between Malcolm Muggeridge, the leader of the British right-to-life movement known as Festival of Light, and Archbishop Anthony Bloom of the Russian Orthodox Church had the following comments:

> **Muggeridge:** I think this horror of pain is a rather low instinct and if I think of human beings I've known and of my own life, such as it is, I can't recall any case of pain which didn't, on the whole, enrich life.
>
> **Bloom:** I think it always does enrich life, and people who try to escape it from cowardice miss something extremely precious.[5]

Anyone who has worked with cancer or AIDS patients in pain realizes what nonsense it is to say that excruciating pain from a terminal illness has any redeeming value. This kind of argument from the opponents of euthanasia only strengthens the convictions of Hemlock Society supporters.

Eastern Religions

In Hinduism, the code of ethics for human conduct has always been flexible and subject to local interpretation.[6] No universally accepted scripture guides ethical decisions. However, the concepts of reincarnation, dharma, and karma would be involved in discussions of euthanasia or assisted suicide. Dharma is defined as duty, righteousness, or religion. The highest dharma is to recognize the Truth in one's own heart. Karma corresponds to the law of cause and effect, manifests as one's destiny, which is caused by past actions, including those of previous lives.

Cutting short someone's life by euthanasia or assisted suicide would be viewed as contrary to one's dharma by interfering with the working out of the karma in one's life. So euthanasia or assisted suicide would create negative karma for both the patient and the physician.

Buddhist doctrine also does not address euthanasia.[7] However, according to the first of the ten precepts of Buddhism, a physician should relieve the suffering of the terminally ill, but not interfere with the working out of one's karmic patterns. Buddhism advocates hospice care, not euthanasia.

CHAPTER 6

The Campaign to Legalize Euthanasia

Derek Humphry became committed to legalizing euthanasia for the terminally ill as a result of the experience of caring for his first wife, Jean, during the last years of her battle with metastatic breast cancer. He ultimately aided her suicide and wrote *Jean's Way,*[1] a moving account of their 22-year marriage and Jean's illness and death in 1975. He wrote the book with the encouragement and assistance of his second wife, Ann Wickett, to challenge the laws against "mercy killing" and to present euthanasia in a rational, positive light.

JEAN'S WAY

At the time of his first wife's terminal illness, Derek Humphry worked as the home affairs reporter for the *Sunday Times* in London, England. Mr. and Ms. Humphry were intelligent, cosmopolitan, politically active, dynamic individuals who forged a strong, enduring marriage partnership. They successfully raised three sons, the last one adopted, in London, England.

Jean had a strong family history of cancer and developed cancer of the breast in her early forties. She underwent mastectomy, radiation therapy, hormonal therapy, and chemotherapy at the Churchill Hospital in Oxford. At the time of Jean's illness, no hospice had been established in the vicinity of Oxford. Humphry's book did not mention any efforts to seek hospice care elsewhere.

She courageously endured pain and disability during remissions and relapses for two years and four months. Finally, she felt that there was no point to living further with the misery of the disease. With the help of her husband, she took a lethal overdose of a sedative and painkiller.

The authorities found out about the assisted suicide of Jean Humphry only when her husband's book, *Jean's Way,* was published. Because of the publicity that now attended the case, the public prosecutor questioned Derek Humphry about Jean's death. In England, assisting suicide is a crime punishable by up to 14 years in prison. Given the extenuating circumstances, the prosecutor decided not to charge Humphry.

From the layperson's point of view, this story appears to present a powerful argument for legalizing euthanasia and assisted suicide for terminally ill people. However, my personal view is that better medical management of the pain and other symptoms of Jean Humphry's cancer, coupled to improved psychosocial support, would have completely changed the course of her terminal illness. Thus, I believe that, had she been given modern cancer pain therapy within the context of a hospice approach to her terminal care, she would not have wanted euthanasia. A careful reading of *Jean's Way* provides lots of evidence to support this opinion.

According to Derek Humphry's account, communication with Jean's physicians had been poor throughout the course of her illness. At the time of her first relapse with painful cancerous metastases to her spine, he was moved to write, "The doctors revealed nothing to Jean about her condition and, while her spirits were good, she complained to me that the medical staff became evasive whenever she enquired about the nature of her illness. Their reticence caused her to suspect that something was seriously wrong and I found it hard to allay her anxiety because I simply did not know the facts or theories which the doctors were considering. Certainly a doctor's reluctance to be honest with a fatally ill patient is justifiable if he feels that frankness will do irreparable harm to the patient's peace of mind, but in Jean's case absolutely no effort was made to assess her ability to cope with the truth. I found it incredible that there were no skilled ancillary workers to assess the patient's emotional needs at such a crucial point. I found it even more astonishing that not one member of the medical staff, either at Southend or Oxford, expressed a desire to talk to me despite the fact that I was always visible and available to them during scores of hospital visits."

When Mr. Humphry finally confronted his wife's doctor about her condition, the reply was frank without expressing emotional support or reassurance about the management of pain or other symptoms. "After a long pause, the doctor weighed her words extremely carefully, 'It looks as if the cancer has come back. We don't know where it is or what part of the body is affected because the pain seems to be traveling. But I should warn you that the situation is very serious and you should prepare your self to expect the worst.'"

Some time later, when the disease became far advanced, Derek and Jean Humphry again wanted information about the medical situation. Cancer specialists at Churchill Hospital in Oxford had run a battery of X-rays and tests, but left it to Jean's general practitioner (GP) to convey the results of the tests. The GP did not discuss the test results or the prognosis with Jean, but gave the news privately to Derek in the following manner, "It was the dreaded moment. The truth would be mine, whether I liked it or not. We walked down the drive towards his car and I sensed that he was gathering his wits in order to say something crucial. When he spoke, his words were not entirely unexpected, although nonetheless they provided me with quite a blow. 'I've had the results from Oxford,' he said quietly. 'The cancer has spread to the bone and you must expect your wife to be dead by the end of the year.'" The doctor left Humphry to convey this news to his wife.

* * * * *

*No specific death sentence should ever
be given to a cancer patient
because no one really knows how long
anyone will live with a cancer.*

* * * * *

The physicians made many errors in the manner they told Jean and her family about her illness and the outlook. The cancer specialists ought not to have delayed telling her, thus passing the buck to the GP. The GP should have told Jean directly, along with Derek and other immediate family members. No specific death sentence should ever be given to a cancer patient, because no one really knows how long anyone will live with a cancer. Jean outlived the doctors' pronouncement by several months. When giving bad news, doctors should realize that patients and family members are quite vulnerable, and need to be reassured that everything will be done to control pain and other symptoms.

Jean's GP could not provide this reassurance, since her doctors had simply not been trained in providing the hospice approach to pain and symptom control and in the optimal psychosocial management of the terminally ill.

Describing one particularly gruesome incident, Humphry wrote, "...areas of pain flared up in Jean's arm, leg, and back. There was no doubt that immediate hospitalization was necessary and when the ambulance arrived Jean knew she was in for a nightmare ride and swallowed as many painkillers as she dared. However, the drugs could not entirely alleviate the discomfort and by the time we reached the hospital, she was crying out in distress. The only alleviation for Jean's condition was radiotherapy on the actual pain-spots, but it was necessary to pinpoint these accurately by X-ray beforehand. Additional painkillers would have caused her to lose consciousness, thus preventing the radiographers from properly positioning her for their series of pictures. However, Jean became so convulsed with pain that she ceased having any control over her own movements; in just a few moments, she became so weak and distraught that I had to prop her up in front of the machine in different positions. The second the radiographers had finished, a doctor who had stood throughout poised with a syringe mercifully injected her with Pethedine (meperidine) and she passed out instantly."

From Humphry's description, the medical management of Jean's pain was no better than the accompanying psychological and emotional support. He described Jean as drugged into unconsciousness so that their communication was cut off. During the last two years of her illness, when the tumor had spread to her bones, she suffered continual, virtually unrelieved pain. In the book he often refers to painkillers and sedatives, but only mentions Pethedine by name. In England, Pethedine is the brand name for meperidine, marketed as Demerol in the US.

The use of meperidine to control the chronic pain of cancer connotes a lack of understanding on the physician's part about cancer pain control. Morphine or hydromorphone would have been more appropriate. Withholding analgesics until after X-rays have been obtained makes no sense. Another mistake was to allow pain to recur before each dose of painkiller rather than giving regular doses to prevent the re-emergence of pain. Jean's doctors obviously were inadequately trained in pain management and palliative care techniques.

Humphry described the effect of Jean's suffering, "Apart from not wanting her to suffer unnecessarily, for I was almost as ravaged as she from the preceding weeks' agony..." A hospice team would have provided support and respite for him and the rest of the family.

If Jean Humphry had been enrolled in a home hospice program, the control of her pain would surely have been far better throughout the course of her disease. With better pain and symptom control, and effective psychosocial support, I personally doubt she would have chosen suicide.

Shortly after Jean's death, Dr. Robert Twycross opened Sir Michael Sobel House, an inpatient hospice, adjacent to the Churchill Hospital in Oxford. Dr. Twycross is one of the world's foremost authorities on the treatment of cancer pain. I began my studies of hospice there in 1979 and returned for a course on hospice in 1985. In America I have not yet seen anything close to the quality of hospice care that I saw at Sobel House. Oxford GPs and cancer consultants are now quite familiar with the hospice approach and do not hesitate to refer patients like Jean for palliative care, either on an inpatient or outpatient basis.

In 1980, Derek Humphry founded the National Hemlock Society (NHS), a euthanasia advocacy group in Los Angeles, California. The NHS now has over 40,000 members nationwide, with over 75 local chapters. It distributes pro-euthanasia books and pamphlets, holds conferences, and promotes euthanasia legalization initiatives. The NHS came into national prominence with the publication of Humphrey's book *Final Exit* in 1991.[2]

FINAL EXIT

The Practicalities of Self-Deliverance and Assisted Suicide for the Dying

Derek Humphry's book, *Final Exit,* attempts to justify "rational suicide" and physician-performed euthanasia of terminally ill people who request it. The book assumes that the reader agrees with the legalization of euthanasia and assisted suicide, and gets down to the practicalities of carrying them out.

Humphry details what are often bizarre and gruesome ways to commit suicide, such as cyanide poisoning, starvation, asphyxiation by plastic bag, and burning to death. He recommends drug overdoses and goes into great detail about which drugs

* * * * *

*Given better physician and nurse training
and the allocation of more resources
to hospice medicine, pain and other
symptoms could be much better controlled
in all terminally ill patients.*

* * * * *

are most effective, which should be avoided, and how to acquire potentially lethal drugs. He also discusses who to tell about committing suicide, letters to be written, insurance issues, double suicides, and support groups for the dying.

Without citing specific references, Humphry states that about 10 percent of cancer patients have pain that cannot be controlled, and claims that uncontrolled pain is a legitimate reason for euthanasia or assisted suicide. Although he gave lip service to hospice doctors who are skilled in pain management, he did not acknowledge that his first wife, Jean, had suffered much unnecessary pain precisely because her physicians lacked knowledge about contemporary pain treatments and techniques.

Even with the poor state of our current care for the terminally ill, it surprises me how few patients with poorly controlled pain from cancer or AIDS attempt suicide or ask for euthanasia. Given better physician and nurse training and the allocation of more resources to hospice medicine, pain and other symptoms could be much better controlled in all terminally ill patients. Just how much better remains to be seen once we acknowledge the problem and begin to work on it.

Humphry suggested that quality of life issues, personal dignity, and loss of autonomy may also lead people to consider suicide and euthanasia. He obviously had not been exposed to the notion that hospice is in fact all about the terminally ill patient's quality of life, personal dignity, and maintenance of autonomy.

Humphry's brief discussion of "The Hospice Option" also contained a regrettable "Catch-22." He states that whether you choose hospice care should depend on how well you and your family cope with your terminal illness. According to Humphry, if you want to keep your options open to commit suicide should your suffering become unbearable, then you should reject hospice care. He did not appreciate that hospice care at its best obliterates unbearable suffering and affords a dignified death with well-controlled symptoms.

POLLS REGARDING EUTHANASIA AND LIFE-SUPPORT TERMINATION

The NHS commissioned the Roper Organization of New York City to conduct a poll of 1500 people in the states of California, Oregon, and Washington in May 1991. Respondents said they would

* * * * *

...hospice care at its best obliterates
unbearable suffering and affords a dignified
death with well-controlled symptoms.

* * * * *

vote to legalize physician "aid-in-dying" with a lethal injection by 60% versus 32%. Eight percent of
the people polled didn't know. Californians were
slightly more in favor than Oregonians and Washingtonians of legalizing physician aid-in-dying. By
an almost three-to-one ratio, those polled favored
physician-assisted suicide for the terminally ill (68%-
yes, 23%-no, 8%-don't know). Of the Protestants
polled, 67% said "yes," as did 66% of the Catholics.

The younger the person, the more likely he/she
would be to agree that euthanasia and physician-
assisted suicide were acceptable practices. Men were
slightly more in favor than were women (3% difference). The better educated and higher income people
favor euthanasia slightly more than those with less
education and lower income.

The journal *Physician's Management* reported
on a poll of 498 doctors conducted in May 1991.
Only 88% said that they would honor a patient or
relative's request to discontinue life support. This
surprised me, since a terminally ill patient—or that
person's proxy—has the legal right to be allowed to
die by foregoing or discontinuing heroic treatment.

If an AIDS patient asked for a prescription for
a lethal overdose of drugs, 10.7% of the physicians

polled indicated they would write it. Another 9.4% said that at some time in their careers they had taken an action resulting in a patient's death.

The Washington State Medical Association (WSMA) conducted a survey of its members on the provisions of Initiative 119 in February 1991. Out of 2000 randomly selected questionnaires sent, 1105 physicians responded (55.25%). Sixty percent of these opposed allowing physicians the legal right to assist suicide by prescribing a lethal injection. Seventy percent opposed granting physicians the right to give a lethal overdose of medication to a terminally ill patient who requests it. By 51 to 49%, they voted to have the WSMA oppose Initiative 119.

A poll taken in March 1992 showed Californians favor legalizing Euthanasia by 75% to 25%. As in Washington state, I expect this to change dramatically during the election campaign.

WASHINGTON STATE "DEATH WITH DIGNITY" INITIATIVE

Based on these and other polls, the NHS felt that a euthanasia ballot initiative was winnable. Finally, Humphry's writing, speaking, and organizing yielded the first public referendum concerning euthanasia in history. The Hemlock Society of Washington state and supporting organizations collected 223,000 signatures to put the legalization of euthanasia to an historic vote on the November 5, 1991 ballot. The "Death with Dignity Initiative" in Washington state was closely followed throughout the country.

Initiative 119—The Adversaries

Proponents of euthanasia and Initiative 119 in Washington saw themselves as modern thinking, humanistic, intellectual people who wanted the ultimate control over their lives. Washington Citizens for Death with Dignity, the main campaign organization, and Washington Hemlock PAC raised the majority of the funds to support Initiative 119.

A long list of Washington state organizations supporting euthanasia includes the American Civil Liberties Union of Washington, ACT/UP-Seattle (and other AIDS support groups), Gray Panthers of Washington, Humanists of Washington, Libertarian Party of Washington, National Association of Social Workers—Washington, National Organization for Women, Washington, Unitarian Universalist Association, Washington, Democratic Party, and Washington state Nursing Home Resident Councils.

People opposing euthanasia often represented orthodox religious organizations. The Washington state Catholic Conference and Human Life Political Action Committee (PAC), an antiabortion group, were the major fundraising organizations opposing Initiative 119. The Washington Medical Association also opposed it. "No on Initiative 119" organizations raised about $800,000 that was used primarily for an effective media campaign.

Dr. Robin Bernhoft, an Everett, Washington surgeon, headed Washington Physicians Against 119. "We all want death with dignity," Bernhoft said.

☛ ☛ ☛ ☛ ☛

We all want death with dignity...
but a doctor killing a patient
with a lethal injection is acting out
abandonment and despair—not dignity.

☛ ☛ ☛ ☛ ☛

"But a doctor killing a patient with a lethal injection is acting out abandonment and despair—not dignity." A nurses organization also formed to oppose 119. Hospice organizations in Washington state and elsewhere generally opposed the measure.

The dramatic turnaround in public opinion in Washington state resulted from an effective "No on 119" media campaign, primarily on television. The issues stressed were the need for effective pain control and effective treatment of depression in the terminally ill. The anti-euthanasia coalition also raised the question: Is it possible that the economic discrimination associated with problems of access to health care will eventually lead to forced euthanasia for minorities and the poor?

Dr. C. Everett Koop, the former Surgeon General of the United States, convinced many Washingtonians to vote against euthanasia in a persuasive television ad. Other trusted figures arguing against euthanasia included terminally ill patients and hospice nurses.

Dr. Jack Kevorkian's double "assisted suicide" of nonterminally ill women in Michigan weeks be-

* * * * *

Is it possible that the economic discrimination associated with problems of access to health care will eventually lead to forced euthanasia for minorities and the poor?

* * * * *

fore the election definitely hurt the pro-euthanasia forces. Many, including Derek Humphry, distanced themselves from Dr. Kevorkian, who was already becoming known as "Dr. Death." The suicide of Derek Humphry's second wife, Ann Wickett, also boosted the support of the "No on 119" coalition. Ms. Wickett was a breast cancer survivor with no evidence of recurrent disease when she overdosed on medication in a fit of depression.

Since polls before the Initiative 119 campaign showed Washingtonians favoring euthanasia by an almost three-to-one ratio, the defeat of 119 by 54% to 46% reflected a remarkable success in public education by the "No on 119" coalition. It remains to be seen whether tangible improvements in the education of medical professionals in palliative care services occur as a result of the campaign.

CALIFORNIA EUTHANASIA INITIATIVE "TERMINAL ILLNESS. ASSISTANCE IN DYING"

Since the National Hemlock Society (NHS), as a nonprofit educational society, cannot spend money on political campaigns, such as those to legalize euthanasia, NHS members decided to form a separate political action committee (PAC). In 1986 two Los Angeles lawyers, Robert Risley and Michael White, working with the NHS, formed Americans Against Human Suffering. AAHS serves as the political arm of the NHS and promotes ballot initiatives and bills in legislatures. In the Appendices, I have included some solicitation letters from AAHS along with my analyses of their arguments.

In 1986 Robert Risley and the Americans Against Human Suffering (AAHS) proposed "The Humane and Dignified Death Act" to the US Congress. If enacted, this resolution would have legalized euthanasia by physicians of terminally ill adults who requested it. When this proposal failed to attract congressional supporters, the AAHS and NHS launched a ballot initiative in California by the same name. After the petition campaign failed to qualify for the 1988 California general election, Derek Humphry moved the NHS headquarters to Eugene, Oregon where he felt the political and social climate would be more favorable to efforts directed to the legalization of euthanasia.

AAHS then launched a new petition drive to legalize euthanasia and assisted suicide in California in the Fall of 1991 to qualify for the November 1992 election. The initiative, titled "Terminal Illness. Assistance in Dying" (*see* Appendix I) supposedly offered more safeguards than the Washington state initiative. Profits from Derek Humphry's best selling book, *Final Exit* (more than 520,000 hard-back copies sold), helped euthanasia supporters collect signatures. As I'm writing this, AAHS has just succeeded in gathering 428,000 signatures (385,000 are required to place an issue before the voters).

In California, a coalition to defeat the initiative was formed in February 1992. Its principal members are representatives of the California Medical Association, the California Nurses Association, the California Commission on Aging, the Catholic church, and other groups. An estimated $3–5 million is being raised to defeat the initiative. Notably, this coalition vowed not only to defeat the euthanasia initiative, but to offer tangible proposals to improve the status of palliative care in California. Members agreed to pursue the latter effort even if the euthanasia initiative did not qualify for the ballot.

The Beverly Hills Bar Association voted to support voluntary aid-in-dying as defined in this and other initiatives. The American Bar Association responded with a resolution opposing euthanasia or assisted suicide.

OTHER EFFORTS TO LEGALIZE EUTHANASIA AND ASSISTED SUICIDE

The legalization of euthanasia has been debated and rejected several times in the English Houses of Parliament. People involved with the English hospice movement have argued forcefully and successfully against "aid-in-dying." They offer convincing arguments that hospice care provides a viable alternative to euthanasia.

The Right to Die by Derek Humphry and Ann Wickett[3] reviews the chronology of the euthanasia movement in the United States over the past 70 years. Before the 1980s, euthanasia advocates made little progress in Congress, in state legislatures, or with ballot initiatives.

Efforts to legalize euthanasia have also been initiated in Oregon. Senator Frank Roberts introduced a NHS-supported bill into the Oregon legislature in 1990. This bill to legalize voluntary active euthanasia and another "right to die" bill failed in the House Family Justice committee. The NHS now plans to gather the 65,000 signatures needed to qualify an euthanasia initiative for the 1994 ballot.

WHICH DOCTORS WILL PERFORM EUTHANASIA?

If euthanasia and assisted suicide were legalized, the practical problem would arise of precisely which physicians would be willing to administer it. At a 1988 conference I attended with about 40 other cancer specialists at the USC–Norris Cancer Hospital in Los Angeles, California, Dr. Westley Robb, a professor of ethics at USC, asked for a show of hands regarding who favored legalizing euthanasia and who would be willing to carry it out if it were legal. Not one hand went up.

If the primary treating physicians of terminal cancer patients are not willing to perform euthanasia, the dying patient would need to find a new physician willing to perform the act. I doubt whether any hospice physicians would intentionally kill patients, even if the practice were legalized. As specialists in pain and symptom management, they realize that terminally ill people do not choose to hasten death unless their pain or other symptoms are insufficiently controlled or the psychosocial support they also need is somehow inadequate. They also understand that with appropriate care—using the hospice team approach—these patients will no longer want euthanasia or suicide.

Despite polls of physicians commissioned by the NHS and others, I doubt whether many doctors of any kind will be available for this "service." In the majority of cases, terminally ill people will need to change physicians in order to find one willing to give a lethal overdose.

Certainly physicians willing to administer euthanasia are unlikely also to be either treating cancer specialists or hospice doctors. Generalists with little training in cancer treatment or hospice medicine might, however, step in to do the job for those few patients who insist. Nonetheless, I predict it will be extremely difficult to find many physicians who will administer euthanasia, considering the emotional burden, the disapproval of major medical associations, the potential damage to a medical practice from the publicity, and the associated social stigma.

CHAPTER 7

Euthanasia in the Netherlands

Dutch court cases, beginning in 1973, led to the current status of euthanasia remaining illegal, but rarely prosecuted. Euthanasia and assisted suicide remain crimes in the Netherlands.

In 1982, Queen Beatrix established the State Commission on Euthanasia to formulate recommendations about legislative reforms relating to the regulation of euthanasia and assisted suicide. The commission subsequently recommended permitting euthanasia under certain conditions, but could not agree on the specifics. Two of the 15-member commission issued a minority report that opposed legalization of euthanasia and assisted suicide.

No fewer than seven other Dutch governmental or professional agencies have drafted guidelines for the practice of euthanasia. Legislative efforts to legalize euthanasia and assisted suicide have likewise failed to gain widespread consensus. Since a Dutch Supreme Court decision in 1984, physicians may be given a waiver of prosecution if ten specific criteria are fulfilled and the case is reported to the Dutch Department of Justice for review. In their decisions, the Dutch Supreme Court and the

Hague Court of Appeals themselves appealed for professional and legislative guidance on the issue. Few in Holland believe that euthanasia or assisted suicide will be formally legalized soon.

The Royal Dutch Society for the Promotion of Medicine (KNMG) offered a series of guidelines for the practice of euthanasia. It supported legalization of euthanasia as a last resort, to be considered only in exceptional or extreme cases. The KNMG-proposed guidelines for the regulation and control of the practice of euthanasia are as follows:

1. Voluntariness on the patient's part
2. A well-considered request
3. Stability of desire
4. Unacceptable suffering
5. Collegial consultation

These proposed guidelines were an attempt to ensure that unwilling patients would never undergo a euthanasia coerced to any degree by their overburdened caregivers. They sought to exclude incompetent patients, depressed patients, or others reacting to situations that may later resolve or improve. Unacceptable suffering was not further defined, but left to the physician's discretion to determine. The collegial consultation guideline does not call for an objective third party, but suggests review with another physician familiar with the case or with clerics, psychologists, or other third parties.

The guidelines leave much open to professional judgment and assume the good intentions and profes-

* * * * *

...guidelines leave much open to professional judgment and assume the good intentions... of practitioners....it is commonly acknowledged that euthanasia... seldom operates under any real guidelines.

* * * * *

sional competence of practitioners. However, it is commonly acknowledged that euthanasia, where practiced, seldom operates under any real guidelines.

Because of problems in documenting the extent of the practice of euthanasia in Holland, frequency estimates have varied between 3000 in a 1985 survey of general practitioners to 12,000 in reports from euthanasia advocates. Physicians report only about 200 cases of euthanasia to the public prosecutors per year.

In 1986 there were 12 cases of suspected abuse of euthanasia. In the first half of 1987, the number rose to 24.

DR. ELSE BORST-EILERS' EXPERIENCE OF EUTHANASIA IN HOLLAND

In December 1989, I attended a public discussion on the euthanasia debate featuring Dr. Borst-Eilers, a euthanasia proponent from the Netherlands. She practiced hematology before becoming an administrator in the University Hospital of Utrecht, a large Dutch city. Her involvement in the euthanasia issue began in the early 1980s when she first heard sporadic rumors about euthanasias in her hospital. Dr. Borst-Eilers felt that, in the interests of all concerned, a formal policy should be adopted for the practice of euthanasia. In 1984 her hospital established a set of guidelines at about the time the Royal Dutch Medical Society proposed that official euthanasia guidelines be enacted into law.

Dr. Borst-Eilers noted that problems arose when doctors practiced euthanasia surreptitiously. The nursing and other professional staff tended to spread inaccurate or misleading stories about the euthanasia cases. No review was possible about the appropriateness of acts of "mercy killing." The hospital and its staff may potentially have been in considerable legal jeopardy by allowing this unregulated practice.

A committee headed by Dr. Borst-Eilers proposed a set of guidelines for voluntary active euthanasia that included the following:

1. Euthanasia would be available as one possible component of an overall policy of practicing good pain control and terminal care.
2. Only voluntary euthanasia would be practiced.
3. Patients must be mentally competent to request it. No proxies are allowed.
4. No children may be included. Euthanasia candidates must be at least 18 years old.
5. Patients need not be terminally ill (have a prognosis of less than 6 months), but must be chronically medically ill with unrelievable pain or emotional suffering.

6. The request for euthanasia must be well considered and consistent over a significant period of time.

7. The patient's doctor must bring in a second consulting physician on the case to determine whether euthanasia is appropriate. This second physician must review the case, but does not actually have to see the patient.

8. On the day before euthanasia is to be carried out, a ward staff conference should be conducted to discuss the case and to answer any questions. This is to prevent the circulation of unfounded rumors and, one hopes, to maintain the good morale of the staff.

9. Only a doctor may perform euthanasia. The task may not be delegated to a nurse or other staff person.

10. After euthanasia is performed, the local public prosecutor's office must be called by the physician and the case reported as an "unnatural death." A written account of the circumstances by the physician must be forwarded within a few days.

The district attorneys in Holland prosecuted a few doctors in some of the early cases of euthanasia, but none were found guilty. In recent years, no doctor has been indicted for murder. Still, officially, the doctor performing euthanasia is the subject of a criminal investigation for murder for about six months until the prosecutor closes the case. This is clearly quite an uncomfortable situation for any physician, and may partly explain why only about 200 of the 3000+ euthanasia cases have been properly reported. Most of the unreported cases occur in the patient's home, where regulation and monitoring are virtually non-existent.

A study of euthanasia in the Netherlands revealed that about 80% of cases involved patients terminally ill with advanced cancer. In another 10% the patient had a severe neurological disease such as multiple sclerosis or amyotrophic lateral sclerosis. These diseases in the advanced stages may lead to total paralysis and dependence on respirators. The remainder of euthanasia cases fell into the "miscellaneous" category.

I asked Dr. Borst-Eilers about the availability of experts in pain and symptom management in her country. She responded that the patient's primary care physician (usually an internist or general practitioner) was the one to manage medical care until the end. If euthanasia were requested, the patient would ask the primary care doctor. She did not know of any hospice programs or physicians specializing in palliative care in the Netherlands.

Dr. Borst-Eilers cited an example of the standard of pain management in her university hospital. Nurses give patients with severe cancer pain bedside bottles containing morphine. Doctors then instruct the patients about the drug and tell them to take as much as they need to control pain.

I wish cancer pain control were that easy. This rather simple technique has many pitfalls and is not considered an effective way to control most cases of cancer pain. According to Dr. Borst-Eilers' account, the average Dutch doctor knew little about pain and symptom management and the overall hospice approach to people with terminal illness.

I concluded from this presentation and discussion that euthanasia is popular in the Netherlands largely because there is no awareness of the alternative—good hospice care.

REGULATING DEATH

Dr. Carlos Gomez published *Regulating Death* in 1991, describing his study of euthanasia in the Netherlands. Dr. Gomez had graduated from the University of Virginia School of Medicine and received his PhD in public policy studies from the University of Chicago. He is currently a resident in Internal Medicine at the University of Virginia.

In January of 1989, Dr. Gomez undertook an interesting study of the practice of euthanasia in the Netherlands. Working from the Free University of Amsterdam, he contacted physicians, government officials, and others involved in the euthanasia debate. He asked the physicians' professional organizations—the Royal Dutch Society for the Promotion of Medicine (KNMG) and the Netherlands Society of Voluntary Euthanasia (NNVE)—for assistance in finding physicians willing to discuss their own practice of euthanasia. He received very little help from these two pro-euthanasia organizations. He also wrote or phoned other organizations, as well as physicians, lawyers, and health officials.

Through his contacts, he was referred to 16 physicians, seven of whom agreed to be interviewed.

He also found two nonmedical people who provided case histories.

By interviewing willing physicians and others, Dr. Gomez documented case histories on 26 euthanized patients. Included in this sample was a two-day-old infant with Down's syndrome and a gastrointestinal tract blockage the parents declined to have treated. An automobile accident victim, who was not expected to survive, also received a lethal overdose of medication from his emergency room physician. One obviously demented woman was euthanized, although she could not give consent.

Nine patients had cancers requiring opioids for pain control. No details of the pain control regimes were given. Of the remaining patients requesting euthanasia, six had strokes, four suffered from heart failure, and two had AIDS. Only four of the cases were reported to the public prosecutor, in keeping with estimates that 4–6% of euthanasia cases have been reported to the authorities.

In this small sample of cases in which the physicians agreed to discuss patients they had euthanized, many violations of the proposed guidelines were noted. In addition to practicing involuntary euthanasia, instances existed of neglecting to consult with colleagues. We have to depend on the opinion of the physicians regarding their patients' "unacceptable suffering." No physician requested the assistance of palliative care specialists to help manage pain or other symptoms, presumably because such expertise does not exist in the Netherlands.

* * * * *

He cited...a case in which euthanasia by lethal injection was attempted, but the patient didn't die. After the patient woke up, he chose to go home rather than undergo another attempt at euthanasia.

* * * * *

In addition to the obvious violations of the guidelines in Dr. Gomez' cases, there were also subtle problems, such as the matter of free choice as opposed to physician discretion. However, voluntary consent necessitates a well-informed patient with adequate knowledge of the facts of his/her case. The "facts" of individual cases depend on the physician's knowledge, skill, experience, and particular habits of practice. They also rest on how a physician chooses to interpret and present the facts of the case, so that patients' perceptions of the reality of their situations is surely colored by the doctor–patient relationship.

In Dr. Gomez' study, it was impossible to verify whether a patient's requests for euthanasia were enduring or temporary. He cited Prof. Van der Meer, who referred to a case in which euthanasia by lethal injection was attempted, but the patient didn't die. After the patient woke up, he chose to go home rather than undergo another attempt at euthanasia.

A related issue was "Who should introduce the topic of euthanasia, the patient or the physician?"

Dr. Peter Admiraal, Holland's most notable physician advocate of euthanasia, said "not to broach the possibility of euthanasia is to deny the patient the full range of available options." Dr. Gomez interpreted this as the physician implying to the patient that the situation was hopeless. I agree.

Dr. Gomez concluded that in the Netherlands the public policy of permitting euthanasia according to agreed-upon guidelines failed to accomplish its stated purpose. It neither controls the practice nor prevents abuses. The Dutch have no way to determine the effect of euthanasia on clinical decision making and doctor–patient relationships. He found that in the Netherlands regulation of euthanasia broadened rather than restricted the practice. Prosecution of abuses has been virtually impossible, since the physician controls the evidence presented to the prosecutor.

The regulations called for a second opinion before performing euthanasia. However, this guideline may be ignored or complied with only to prevent legal complications. An unbiased second opinion was not required, since the treating doctor chooses the other physician to evaluate the appropriateness of euthanasia. The second physician does not have to see and examine the patient, but may reach an opinion based only on discussion with the primary doctor.

Dr. Gomez predicted that, compared to the Netherlands, the regulation of euthanasia would be much more difficult in the United States. Citizens in the

* * * * *

The proponents of euthanasia...
do not propose adequate regulatory mecha-
nisms. There is no civilized way
to regulate euthanasia.

* * * * *

Netherlands have universal access to health care services administered by the State. In the United States, the crisis in health care funding and access is increasingly leading to different standards of care for the rich and the poor. This will certainly have direct and indirect influences on perceptions of "voluntary consent" and "intolerable suffering."

Throughout *Regulating Death,* Dr. Gomez never addressed the state-of-the-art of palliative care in Holland. He did not mention whether he had found hospices for terminally ill patients or specialists in pain control and symptom management.

Interestingly, Dr. Gomez remained open to the possibility that euthanasia could be adequately regulated, although he did not know how.

The proponents of euthanasia likewise do not propose adequate regulatory mechanisms. There is no civilized way to regulate euthanasia.

Paul Van Der Maas' Study of Euthanasia and Assisted Suicide in Holland

In September 1991, Dr. Paul J. Van Der Maas and other Dutch researchers published the final results of a nationwide study of euthanasia and medical decisions at the end of life.[3] This article reported three studies looking at the practice of euthanasia. In one study, the authors interviewed 405 physicians from different disciplines about their experience with euthanasia and assisted suicide. These same physicians agreed to report all the euthanasia and assisted suicide deaths in their practices over the subsequent six-month period. The researchers also selected a random sample of 7000 deaths and sent questionnaires to the primary physicians. A summary of the findings of these studies about medical decisions at the end of life included the following points:

1. Physicians, patients, and family formulated medical decisions about the end of life in 54% of all non-acute deaths.
2. Nontreatment decisions (NTD). The withholding or withdrawal of treatment in situations where the treatment would probably have prolonged life occurred in 17.5% of all deaths.
3. Alleviation of pain and symptoms (APS) with opioids in such dosages that the patient's life might have been shortened occurred in 17.5% of all deaths.
4. The physician performed euthanasia at the request of the patient in 1.8% of all deaths.
5. The physician provided medication for the patient to commit suicide in 0.3% of the cases.
6. In 0.8% of deaths, the patient made no explicit or persistent request for euthanasia or physician-assisted suicide, but the physician performed euthanasia anyway.

These data were noteworthy for several reasons. Medical decisions about the end of life should be made for all people with terminal illnesses when death is expected. The fact that in only 54% of expected deaths were end-of-life decisions made suggests that physicians were poorly educated in palliative care.

The alleviation of pain and symptoms (APS) with opioids rarely shortens life (once or twice per 1000 patients in my experience). Indeed, good pain and symptom management frequently prolongs life. The high percentage of patients whose lives may have been shortened by opioids ordered to relieve pain (17.5%) calls into question the training of Dutch physicians in the use of opioids and other analgesics for pain of terminal illness. The failure to adhere to even basic guidelines is demonstrated by the 0.8% of nonvoluntary euthanasia cases.

In conclusion, the Dutch experience with euthanasia provides a frightening precedent. It demonstrates that the absence of the hospice alternative greatly increases the attractiveness of euthanasia as an option. And it demonstrates the difficulty of regulating the practice of euthanasia to avoid abuses even with guidelines in place.

CHAPTER 8

On Pain and Living

> I ask for a natural death
> no teeth on the ground
> no blood about the place
> it is not death I fear
> but unspecified, unlimited pain.
>
> —*Robert Lowell*
> *Death of a Critic*

Alexander Pope, a 17th century poet and essayist said: "No greater good can man attain than to alleviate another's pain." Dr. Charles Moertel, past president of the American Society of Clinical Oncology reiterated Pope's sentiments in context of the 20th century: "Relief of pain is the salient and most overriding responsibility of the physician caring for the patient with incurable cancer." Virtually every oncologist, internist, and family practitioner would agree with this statement, but many have not developed skill in the techniques of pain control.

Americans fear cancer and AIDS more than any other catastrophic event, including nuclear war. Typically, people fear the pain from these diseases more than dying. Someone in severe pain cannot spend quality time with family members or loved

* * * * *

*Pain dominates everything. So despite the
immense popularity of books, articles,
lectures, and courses on death and dying,
most cancer and AIDS patients are far more
concerned with pain and living.*

* * * * *

ones, attend to unfinished business, or prepare psy-
chologically, emotionally, and spiritually for death.
Pain dominates everything. So despite the immense
popularity of books, articles, lectures, and courses
on death and dying, most cancer and AIDS patients
are far more concerned with pain and living.

What is pain? Pain is defined as "an unpleasant sensory and emotional experience associated
with actual or potential tissue damage, or described
in terms of such damage."[1] In patients with advanced cancer, it is impossible to separate the pain
from organic disease or tissue destruction from pain
of psychological origin or social isolation. We need
to address all components of the pain—physical,
psychological, social, and spiritual.

Surveys show that more than 70% of advanced
cancer patients have pain related to their malignancies.[2] In about 10% of cases, patients complain
of only mild to moderate pain. The rest suffer with
severe pain unless adequately treated. More often
than not, pain dominates the course of the illness.

* * * * *

*The pain of cancer is particularly
meaningless, serving no useful purpose, such
as warning the sufferer of imminent harm
...cancer pain usually grows more rather than
less severe....expand[ing] to occupy a
patient's whole attention....
When this occurs, the patient may believe
that life is no longer worth living.*

* * * * *

Much more is known about cancer pain than
pain from AIDS. Few studies have been done to
determine optimal treatment of various AIDS-related pain syndromes.

Studies report that younger cancer patients more
frequently have pain. This may be because younger
people have more physical stamina and battle the
disease longer. The psychological burden would also
appear greater for a younger person who has a
dependent family or a promising career.

Cancer- or AIDS-related pain differs from the
temporary pain caused by accidents or operations
in several ways. Most important, it is generally
continuous and predictable. The pain of cancer is
particularly meaningless, serving no useful purpose,
such as warning the sufferer of imminent harm.
Unless appropriately treated, cancer pain usually
grows more rather than less severe. Without ade-

quate treatment, it frequently expands to occupy a patient's whole attention and isolates him or her from the world around. When this occurs, the patient may believe that life is no longer worth living. One survey reported that 69% of cancer patients would consider committing suicide if their pain was not adequately treated.[3]

Several types of pain may afflict a cancer or AIDS patient. The right treatment for each depends on whether the cancer involves bones, nerves, internal organs, brain, and so forth. A precise diagnosis of the type of pain helps the doctor design specific treatment. X-rays or other diagnostic studies may occasionally be necessary to define the cause of the pain or to plan radiation treatment.

A cancer patient's pain occasionally may not be related to the malignancy. Toothaches, arthritis, sore muscles, and ordinary headaches bother people with malignancies as well as the rest of us.

COMPONENTS OF CANCER PAIN

Regardless of the cause of the pain, British hospice workers teach that there are four components of cancer pain. These four components are:

- *Physical pain* results from cancer stretching, compressing, or otherwise irritating nerve endings in the body.
- *Psychological pain* relates to fear, anxiety, and depression caused by having cancer and having physical pain.

- *Social pain* results from the isolation from family and friends that the patient feels when his attention is always riveted on fighting the physical pain.
- *Spiritual pain* occurs when life has lost its meaning and living is deprived of purpose or value.

Diagnosis and Treatment of Pain

If a patient claims to suffer pain, pain is being suffered, though no test or instrument can accurately assess the subjective feeling of pain. Modern technology is not likely to produce an instrument capable of evaluating the intensity of pain in the foreseeable future. Hospice physicians generally rely on the patient's report and on an examination to diagnose the pain problem, rarely ordering X-rays or scans, since these tests seldom help with the treatment of pain. Human perception and skill are required to adequately diagnose and treat the pain from cancer or AIDS.

Pain is multidimensional, having many components and varying intensities. This explains to some degree why it is so confusing and frustrating to patients, families, doctors, and nurses. Therefore, the treatment of pain is multifaceted.

"The treatment of pain begins with an explanation," says Dr. Robert Twycross, the renowned cancer pain researcher from Sir Michael Sobel Hospice in Oxford. Patients want to know what caused the pain; whether the tumor is advancing; whether the pain will get worse or go away; whether pain means that the overall prognosis is poor, and whether pain medicines can control the pain. Naturally, patients are especially emotional and vulnerable at this time.

Whether pain medicines will work in particular cases depends greatly on the patient's state of mind and confidence in the physician. The physician's attitude and beliefs will inevitably be transmitted nonverbally to the patient and will influence the results of treatment. Unhurried, thorough explanations by the physician can help immensely in promoting a positive attitude in the patient.

DRUGS USED TO CONTROL CANCER PAIN

For mild pain caused by cancer or AIDS, doctors usually prescribe an analgesic such as aspirin or acetaminophen (Tylenol, Datril). For safety reasons, I generally prescribe magnesium choline salicylate (Trilisate) or salsalate (Disalcid). These are much less likely to cause stomach ulcers, bleeding, or kidney problems. A combination of aspirin or acetaminophen with codeine works well for pain of moderate severity. Treatment of severe pain requires opioid (narcotic) analgesics, usually morphine, hydromorphone (Dilaudid), methadone (Dolophine), or fentanyl (Duragesic).

Morphine is one of a score of drugs found in the opium poppy grown in the Middle East. The name morphine comes from Morpheus, the Greek god of sleep and dreams. In 1680, the renowned physician Sydenham wrote: "Among the remedies which it has pleased almighty God to give man to relieve his suffering, none is so universal and so efficacious as opium." Chemists succeeded in separating morphine from the other drugs in the opium plant in 1803. Around 1900 the famous physician, Sir William Osler, referred to morphine as "God's own medicine."

In the past 50 years, chemists have succeeded in synthesizing many other opioids. These researchers always hope to discover an analgesic that effectively controls pain, has few side effects, and does not cause addiction like morphine. They have not succeeded. Morphine remains unsurpassed for severe pain related to cancer.

Personally, I never recommend meperidine (Demerol), a synthetic opiate, for controlling chronic cancer pain. Although it is a powerful narcotic, it lasts for only two to three hours. Frequent injections or tablets would have to be taken, causing pain and inconvenience, especially at night. In high doses over several days, meperidine tends to increase muscular irritability and occasionally causes seizures. Unlike morphine, methadone, and hydromorphone, which give increasing amounts of analgesia when the dose is increased over at least a tenfold range, meperidine provides little or no additional analgesia above its usual dose of 100 mg by injection.

Pentazocine (Talwin), a weak opioid, frequently causes hallucinations and other unpleasant side effects. When used together with morphine or other potent opioids, pentazocine acts to reverse the pain-killing effect. Like meperidine, pentazocine has no place in the treatment of chronic cancer pain.

Methadone (Dolophine) is the equal of morphine in controlling pain, but tends to accumulate in the body. If I give a patient methadone to control pain, I must be more careful in titrating the dose. The advantage of methadone, however, is that it is inexpensive and can be given every six or eight hours, rather than every four hours.

Hydromorphone (Dilaudid) provides a good substitute for morphine in terms of strength and duration of action (about four hours). Since hydromorphone is four times as potent as morphine, we quarter the morphine dose.

Tetrahydrocannabinol (THC), the active chemical in marijuana, is available for medical uses as dronabinol (Marinol). It alleviates the nausea associated with chemotherapy and may soon be proven useful as an appetite stimulant. Anecdotal reports suggest it may be effective with cancer or AIDS pain. No controlled studies of dronabinol as an analgesic have been published or planned. I do not recommend its use to treat pain. Better medications are available.

HEROIN FOR CANCER PAIN

Diacethylmorphine (heroin), derived from morphine, is available for medical purposes in Britain and Belgium, but not in the rest of the world. Physicians of the 1950s and 1960s felt that heroin controlled pain better than morphine and produced less nausea, vomiting, constipation, and sedation. It also seemed to work better to improve appetite, to prevent coughing, and to relieve anxiety in people with trouble breathing.

In 1977, Dr. Robert Twycross, working at Saint Christopher's Hospice, reported a clinical study showing that morphine and heroin given by mouth have equal effectiveness and comparable side effects. Studies in US cancer hospitals confirm his findings.

Compared to morphine, heroin's single definite advantage is higher solubility. This means that more of the drug can be dissolved in a given amount of liquid for injections. This helps when the patient cannot swallow the opioid solution and requires hypodermic injection. Heroin injections themselves will not hurt as much as injections of other opioids.

We still do not need to legalize heroin for such situations because high potency Dilaudid is now available. This drug has an even greater opioid effect per unit of solution than heroin. An injection would rarely require more than a cubic centimeter (1/30 ounce) of liquid.

HEROIN LEGALIZATION BILLS

Since 1983, Senators Daniel Inouye from Hawaii and Dennis DeConcini from New Mexico and Congressman Henry Waxman from California have introduced bills to legalize heroin for use with terminally ill cancer patients. William Buckley praised this effort in his nationally syndicated column. Many well-meaning citizens support this proposed legislation.

The stimulus for the bills to legalize heroin for cancer patients appears to come from the Committee Against Intractable Pain, a lay group that has had friends or family members die of cancer in un-relieved pain. They reason that the availability of heroin accounts for cancer pain being much better controlled in England than in the United States.

The Committee Against Intractable Pain and Congress are to be congratulated for focusing attention on the problem of inadequate control of cancer pain. However, using heroin instead of morphine would not reduce the pain and suffering more than alternative pain medicines; better physician and nurse training in hospice-style pain management would. In short, we need to fund training programs to educate physicians and nurses in this important area.

PRESCRIPTION ACCOUNTABILITY AND PATIENT CARE IMPROVEMENT ACT

Currently, ten states, including California, have a multiple-copy prescription system for monitoring controlled drugs. Theoretically, this operates to inhibit physicians from inappropriately ordering opioid and other controlled drugs. Studies have documented up to a 50% reduction in such prescriptions once a multiple-copy prescription program is instituted. But since they do not reveal which patients receive less controlled medication than before, we don't know how much drug abuse and the illegal diversion of controlled drugs is reduced. Whether the proper use of these drugs in pain and symptom management for the terminally ill will be unwarrantedly reduced remains unclear.

Legislation to create a nationwide triplicate prescription system failed to be enacted by Congress
in 1991. Representative Pete Stark of Sacramento,
California, who proposed this legislation, estimated
that its enactment would save Medicare up to $1
billion per year in reduced cost of medication and
fewer adverse consequences from the abuse of prescription drugs. Rather than reintroduce this bill,
Representative Stark now proposes an electronic
surveilance system to monitor every prescription
nationwide.

Representative Stark's "Prescription Account-
ability and Patient Care Improvement Act of 1992"
supposedly will detect and prevent illegal and inappropriate dispensing of controlled drugs, including opioids, stimulants, and tranquilizers. This act
would require each state to set up a computerized
databank to record each controlled drug prescription. The permanent record made of each physician's
prescribing habits would be used to determine who
might be diverting medication for illegal purposes.

In general, palliative care specialists oppose
Representative Stark's bill because they fear it will
impede physicians from prescribing adequate
amounts of opioids for pain. They reason that most
physicians already grossly underprescribe opioid
for cancer pain patients, so that this bill will only
make the situation worse.

I proposed a potential compromise to Representative Stark that might help enlist the support
of the palliative care community. This recommen-

dation was that he include in his bill provisions to fund cancer pain services in every medical-school teaching hospital in the country. This way, physicians may learn how best to prescribe opioids where they are indicated. To date, my proposal has not been responded to by the Congressman.

MYTHS ABOUT CANCER PAIN

There are six myths about opioids that may result in inadequate pain control:

• **Myth #1**—"Addiction to opioids is a sign of depravity and should be avoided at all costs."

Some doctors avoid prescribing opioids for fear of making addicts of their cancer patients. This is a tragic mistake. The dictionary defines drug addiction as a "compulsion or overpowering drive to take a drug in order to experience its psychological effects." Adequately medicated cancer patients experience neither this compulsion nor an irresistible psychological drive for the next dose. They do not require constantly escalating doses to achieve the same effect. If their pain is continually controlled by the medicine, they do not behave like craving addicts who think of nothing but when and how they will get their next fix.

• **Myth #2**—"If I give you morphine now, no strong pain medicine will be available later when you need it."

Recent research demonstrates that opioids can be used for years to relieve severe pain from can-

* * * * *

Some doctors avoid prescribing opioids for fear of making addicts of their cancer patients. This is a tragic mistake.

* * * * *

cer. There is no reason to avoid strong opioids for severe pain at any point in a cancer patient's illness. I currently follow three cancer patients who have been on opioids for more than five years with good pain control.

• **Myth #3**—"Oral medications don't work for severe cancer pain. Injections are always required."

Some outdated pharmacology texts say that the dose of morphine given orally must be at least six times the dose delivered by injection. These texts are wrong. British and American researchers have conclusively shown that oral opioids can control cancer pain. Patients generally require about three times the dose by mouth that is needed by injection.

Oral pain medication frees the patient from the hospital and from the discomfort of injections. It also allows patients to control their lives in one sphere, giving them a measure of independence.

• **Myth #4**—"Totally relieving cancer pain produces doped zombies unable to think clearly or function normally."

Although patients may become unusually sleepy for two or three days after beginning appropriate doses of oral morphine, they soon return to being alert and mentally sharp. The morphine causes some of the sleepiness. However, sleep deprivation because of the pain itself is the most important factor.

The original British hospice pain medicine, the Brompton cocktail, contained cocaine. Doctors initially included cocaine to counteract the sedative effect of morphine. However, a clinical study by Dr. Twycross revealed that a stimulant such as cocaine is not needed. The patient should instead be told to expect to be sleepier for two or three days after beginning the medication. This almost always resolves after that time. Cocaine and other stimulants can be avoided in most cases.

• **Myth #5**—"Pain medicine should be given as needed (P.R.N.)."

Inexperienced medical staff may stop the opioid medication each time the patient falls asleep or allow the medicine to wear off before the next dose is given. This leads to a recurring cycle of patient agony alternating with sleep or lassitude, rather than continuous pain relief even as alertness is maintained. With the standard opioid medications given as needed (P.R.N.) instead of around the clock, severe pain will recur four to six times per day.

A four-hourly dosage schedule of oral morphine or hydromorphone, or a six-hourly dose of methadone, is appropriate. The newer long-acting morphine preparations, MS Contin and Oromorph, are more convenient oral medications and require a dose only twice a day. This allows the patient to sleep all night without awaking in pain and begging for more opioid medication. Fentanyl (Duragesic) patches applied to the skin of the chest provide 72 hours of continuous analgesia. This works even when the patient cannot swallow and provides an alternative to injections and suppositories in that situation.

Whatever analgesic is used, the dose should be adjusted so that the pain does not come back between doses.

   • **Myth #6**—"High doses of opioids act as a form of euthanasia."

I rarely see physicians prescribe more than 20 milligrams (mg) of morphine each four hours, or a comparable amount of another opioid. Yet the patients at St. Joseph's Hospice in London whose pain is well and humanely controlled average 30 milligrams each four hours and at times are given 100 milligrams or more without ill effect. Patients with pain from cancer or AIDS in my hospital require, on average, 50 mg of morphine every four hours, or the equivalent medication.

Some people mistakenly believe that the toxic side effects of the opioids shorten the lives of cancer patients. No one questions that opioids have side effects. However, with skillful management of opioid analgesics, including anticipation of possible adverse side effects, no evidence exists that these drugs shorten life.

Indeed, hospice physicians think that good pain control may lengthen the cancer patient's life by avoiding the complications of chronic pain. Such complications include poor appetite (anorexia), which leads to weight loss, weakness, and increased risk of infection. Insomnia caused by pain further weakens and exhausts the patient. Depression, often accompanied by immobility, adds to the patient's debility. So poorly managed chronic pain itself may lead to poor nutrition, increased debility, and susceptibility to infection, all of which may hasten the person's death.

The question of the effect of good pain control on the duration of patient survival must of course never be studied scientifically—on good moral grounds. The reason is that it would require a clinical study comparing cancer patients whose pain is treated appropriately with others who are allowed to suffer because their physicians wrongheadedly give inadequate doses of opioids.

SIDE EFFECTS OF OPIOIDS

Like any other medicine, opioids may have side effects. The best way to manage these adverse effects is to anticipate and prevent them.

Depression of breathing causes the most concern, and is the actual cause of death in a fatal opioid overdose. This doesn't happen, however, when the patient is carefully monitored with a gradual increase in opioid dose until complete pain control is achieved. If regular morphine or another opioid that lasts 3–4 hours is used, the patient should even be awakened from sleep to receive opioid medication. More commonly now, long-acting morphine, (MS Contin and Oramorph) or fentanyl (Duragesic) patches are used for around-the-clock opioid analgesia.

Constipation occurs almost invariably in cancer patients receiving opioids. For this reason laxatives should be given on a daily basis to prevent it. A combination of a stool softener with a mild bowel stimulant works best. Senokot S has the advantage of being a single medication that does the work of two. It contains senna, a mild laxative composed entirely of vegetable extract, and a stool softener. Pericolase works in a similar fashion. Milk of magnesia together with two docusate (Colase or DOSS) stool softener tablets per day is an alternative. A few people need double or triple the standard dose and some require less.

Difficulty with emptying the bladder occurs occasionally in men with enlarged prostate glands who are receiving opioids, especially if anti-anxiety or antidepression medications are also being taken. Adjusting the doses of medications or mechanically draining the bladder with a catheter may alleviate the problem.

Nausea and vomiting are occasional side effects of opioid medications. Antinausea medicines, such as prochlorperizine (Compazine), trimethobenzamide (Tigan), metoclopramide (Reglan), and so on, can usually resolve these problems. I generally prescribe the antinausea medicine as needed when starting the opioid. Nausea and vomiting usually occur only when patients begin on opioids, so the antinausea medication need not be continued indefinitely. Rarely, the patient may need to switch to a different opioid because of uncontrolled nausea and vomiting.

Another annoying side effect of opioids is dry mouth, which requires frequent sips of fluids. This becomes somewhat less bothersome with time.

NON-OPIOID TREATMENTS OF CANCER PAIN

A multifaceted or comprehensive approach is likely to prove most successful in treating pain from cancer. This means that attention must be paid to the physical, psychological, social, and spiritual components of pain. The skillful use of opioids and other medications must be accompanied by good communication and, sometimes, other nondrug pain control techniques.

Lying alone in bed magnifies pain. Visitors distract patients from otherwise dwelling on their miseries. Writing, reading, painting, and working are all good medicines. Sigmund Freud had cancer of the jaw for 16 years, but diverted his attention from his discomfort by writing about psychoanalysis.

For some people, alcohol helps! A cocktail before dinner or wine with meals stimulates the appetite and promotes conversation. Alcohol itself acts as an analgesic, although a weak one. It relaxes and sedates the patient. Both these effects aid opioids in controlling pain.

Occasionally, antianxiety medications or antidepressants must be added to opioids to control

* * * * *

*Radiation therapy and chemotherapy
will help pain if they cause a remission
of the tumor. They should never be relied
upon as the only hope of pain control...*

* * * * *

pain. This makes sense when there is a prominent psychological component to the pain. Mixing these powerful drugs with opioids requires considerable skill and careful observation. The doctor should continually reassess the different medications, always being mindful of side effects, toxicities, and changes in the patient's needs.

Radiation therapy and chemotherapy will help pain if they cause a remission of the tumor. They should never be relied upon as the only hope of pain control, however. The opioid dose can be reduced as these anticancer treatments control the cancer.

In cases in which pain is localized to a single area, anesthetic or surgical procedures can control the pain by injecting or cutting the nerves that forward the pain impulses to the brain. Anesthesiologists and neurosurgeons have tried a variety of procedures designed to help relieve cancer pain; these procedures are often quiet effective.

Acupuncture, hypnosis, exercise, transcutaneous electrical nerve stimulation (TENS), and biofeedback are nondrug techniques used to treat pain of noncancerous origin. They should be investigated as methods of cancer pain control, where they might either supplement drug treatment or be tried alone in certain instances. Research suggests that these techniques may themselves increase the body's own production of endorphins and enkephalins (morphine-like chemicals), and so can help us explore how the body can control pain by itself.

THE PLACEBO RESPONSE

The word placebo comes from ancient Latin and means "I shall please." Today it refers to chemically inactive substances given to patients who are told that it may help their symptoms. Placebos have led to clinical improvement in a wide variety of conditions, including cardiovascular diseases, rheumatoid and degenerative arthritis, gastrointestinal disorders such as peptic ulcers and nausea, migraine headaches, allergies, acne, multiple sclerosis, diabetes, depression, and anxiety. In many studies, placebos have had a significant success rate in treating pain from various diseases.

It is remarkable that most responses to medical treatment prior to the advent of modern day drugs and surgery probably were placebo responses. Yet recognition of this powerful tool has occurred only recently. Placebo did not appear in the title of an article in a medical journal until 1945. Over the centuries, many people have certainly been helped as much by their encounters with physicians as by their medicines; in short, by the placebo the encounters them-

* * * * *

Although placebos may be very powerful,
no responsible physician recommends using
placebos alone to treat cancer pain.
However, part of the art of medicine
is to elicit the placebo response
to supplement the effect of analgesics.

* * * * *

selves afford. I agree with noted author Norman Cousins, who says, "An understanding of the way the placebo works may be one of the most significant developments in medicine in the 20th century."

Although placebos may be very powerful, no responsible physician recommends using placebos alone to treat cancer pain. The placebo response depends on the patient's believing that the medicine will help, and in truth, any medicine will be more effective if a patient and physician expect it to work. Part of the art of hospice medicine is to elicit the placebo response to *supplement* the effects of medications. This remains an important, but underutilized, ally in treating pain from cancer and AIDS.

Unfortunately, the modern emphasis on science and technology in medicine has spawned misconceptions and unhealthy attitudes about placebos. Dr. Goodwin reports that in one medical center 60% of physicians and nurses give patients placebos to see whether pain is "real."[4] Reduction in pain with placebos means to these physicians and nurses that the patient is faking pain or has pain "only in his head."

Dr. Goodwin also notes that 75% of physicians use placebos on problem patients who are disliked or who they feel are exaggerating their pain. "Placebos are used with people you hate, not to make them suffer, but to prove them wrong," reported one senior medical resident.[4] So placebos tend to be reserved for patients for whom conventional treatment has failed or who are deemed undeserving of sympathy. The latter is of course an unworthy, deplorable practice. More enlightened physicians will recognize that a lessening of pain occurring when taking a placebo does not mean that the pain was imaginary.

Placebos also serve to help release the frustrations of the medical staff. Dr. Pepper expressed a similar attitude in 1945: "The giving of a placebo...seems to be a function...which like certain functions of the body, is not to be mentioned in polite society."[5]

To combat the widespread ignorance and misunderstanding about placebos, Dr. Beecher reviewed many scientific studies on cancer and pointed out several interesting facts.[6] About 35% of people in pain get at least some relief from placebos. But morphine is more effective in controlling pain in those who respond to placebos. Placebos are more effective in controlling pain in situations of high stress than at other times. Placebos can have side effects, especially involving the nervous system.

Response to placebos does not imply gullibility or simple-mindedness. Intelligent and sophisticated people are as responsive to placebos as other people. According to Parkhouse, "The subject's response to placebos is more dependent on the particular circumstances surrounding the placebo administration than on the subject's personality....Given the proper circumstances anyone could be a placebo responder."[7]

HENRY AND THE MIRACULOUS TENS

A transcutaneous electrical nerve stimulator (TENS) is a small, battery-operated device that sends a weak electrical current through the skin overlying the area of pain. Its mechanism of action remains uncertain, but it seems to work either by releasing the body's own endorphins or by distracting the patient's attention from the pain. Several years ago, I tried a TENS device on Henry, a multiple myeloma (a form of bone cancer) patient with long-standing, poorly controlled pain.

Henry was an intensely independent, somewhat stubborn, and stoical man in his early 60s. He began experiencing chronic back pain seven years previously. Two years elapsed before myeloma was diagnosed. Radiation and chemotherapy led to a clinical remission with good control of his pain for about three years. For more than two years, I had worked with Henry, trying to help with his back pain. Despite my best efforts at explaining the importance of his medication regime, he resisted taking sufficient opioid to continuously and effectively control his pain.

I visited Henry when he was finally hospitalized for back pain unrelieved by high doses of opioids. He couldn't get out of bed or get to a sitting position. He told me that, before he had come to the hospital for the first time, he took all of the prescribed dose of the opioid. However, he still had experienced no relief.

While Henry was in hospital I increased his medication from six to seven hydromorphone tablets every four hours. He said that this helped his pain, but caused hallucinations. We seemed to be at an impasse when I suggested using a TENS device to supplement his opioid medication. He was more than willing to try anything but higher doses of opioids.

The TENS device immediately helped Henry. He was able to get up and walk with assistance by the end of the first day. He left the hospital with his own TENS machine three days later. Over the remaining three months of Henry's life, he used the TENS on a daily basis to supplement his opioid and make his life more comfortable. Whenever possible he would lower the opioid dose to four or five tablets every four hours. He needed to feel a sense of control over his life, including his pain management program. The TENS device helped him satisfy that need.

EFFECTIVENESS OF PAIN CONTROL TECHNIQUES

With a concerted effort involving the routine use of opioids and ancillary medicines and treatments, almost 80% of hospice patients in Britain now live pain-free or suffer only mild pain. They remain alert and often stay at home until they die.

About 20% of hospice patients continue to have moderate pain despite maximum efforts by the hospice staff. However, most of these patients only have pain when engaged in activities such as walking, gardening, or other physical work. Some further modification of lifestyle usually alleviates the pain in these cases. Only about one patient in 100 in British hospices continue to have poorly controlled pain despite all the best efforts of the hospice staff.

PAIN TREATMENT, A LOW PRIORITY IN AMERICAN MEDICINE

America lags far behind Great Britain in dealing with cancer pain. Most American cancer and AIDS specialists and oncology nurses are not taught the hospice techniques for controlling pain. A survey of recent oncology textbooks showed that only 20 pages out of 5000 dealt with managing the pain of cancer.

The National Cancer Institute (NCI) and various charities spend over $1.5 billion yearly on cancer research, education, and treatment. Only a little more than $1 million goes to pain control research. Pain control education programs were given less than $100,000 per year by the NCI in mid-1980s. Despite the American public's perception of the importance of pain control in cancer and AIDS treatment, the medical establishment has not responded with appropriate training and research programs.

* * * * *

...although we possess the best pain-fighting medicine available and command the world's most sophisticated medical technology, American cancer patients often needlessly suffer great pain as they lie dying.

* * * * *

In short, although we possess the best pain-fighting medicine available and command the world's most sophisticated medical technology, American cancer patients often needlessly suffer great pain as they lie dying. The Medical Knowledge Self-Assessment Program, published by the American College of Physicians, reports: "One third of hospitalized patients treated with opioids are undermedicated, usually because of physician misinformation, under-dosage, or exaggerated fears of addiction."

Judging from my experience in many American hospitals, both teaching and private, this is a very conservative estimate.

PAIN CONTROL EDUCATION

When I first went to England to study hospice in 1979, I didn't go to learn pain control. I had no idea that my pain management skills were grossly lacking. *I didn't know that I didn't know how to treat cancer pain* because I had rarely seen it done successfully. Once I observed how well hospice tech-

* * * * *

*...most American physicians remain
as uninformed as I had been
before visiting English hospices.
They simply don't know that they don't know
the optimal pain control procedures.*

* * * * *

niques worked to control pain, I became determined
to practice and teach these pain management techniques to American oncologists, physicians in training, medical students, and nurses.

Unfortunately, most American physicians remain as uninformed as I had been before visiting
English hospices. They simply don't know that they
don't know the optimal pain control procedures. The
medical oncology specialty examination has no questions on cancer pain management.

In order to improve the control of cancer pain
in the US, medical students, interns, residents, and
cancer specialists absolutely must be trained in palliative care. Oncology specialists in training, particularly, should be required to spend several months
practicing in a hospice setting to learn the optimal
techniques of pain and symptom control. I have
helped to arrange hospice experiences in England
for four oncology fellows in training from UCLA
and USC Schools of Medicine. In each case their
careers in cancer therapy were profoundly and posi-
tively affected.

* * * * *

*The goal of a well-formulated cancer pain
control program is to control
all the pain all the time.*

* * * * *

The goal of a well-formulated cancer pain control program is to control *all* the pain *all* the time.
Oncologists and other physicians caring for cancer
patients should always strive for this goal. If a
physician cannot control a cancer patient's pain,
he or she should seek help from those experienced
with hospice pain management techniques.

WHAT TO DO FOR POORLY CONTROLLED CANCER PAIN

What should you do if you or your loved one
has uncontrolled pain from cancer?

The overwhelming likelihood is that better pain
control can be achieved. Dr. Twycross and other
hospice physicians recommend seeking a second
medical opinion whenever encountering difficult pain
control problems. Often a different slant or fresh
approach to such a problem will reveal a better
solution.

Hospice professionals and anesthesiology pain
specialists can help oncologists and other physicians manage their patients with pain from cancer
or AIDS. However, the patients' physicians must

ask for the help. Often the patient or family member must ask the physician to request consultation from a hospice physician or pain consultant. The reference list for this chapter contains articles that may help you and your doctor better understand the treatment of pain from cancer.[8–10]

CHAPTER 9

Hospice Care
and Standard Oncology

*The word hospice can be traced back to a medi-
eval French term for "inn for weary travelers."
The modern hospice movement began in England
in the 1960s.*

DEVELOPMENT

Cicely Saunders interrupted her university studies during World War II to become a nurse. She
wanted to help her country in its time of need.
Because of a back injury, she subsequently gave
up nursing, and went into social work. In 1948,
while at St. Luke's Hospital, in Bayswater, England,
she tried to help arrange care for an unfortunate
Polish refugee who was suffering a painful death
from cancer. Largely because of this experience,
she resolved to improve the care of dying cancer
patients and became a physician in the 1950s.

Dr. Saunders spent several years at St. Joseph's
Hospice serving the "cockneys" of a poor area of
central London. In 1968, she opened St. Christopher's

Hospice, England's premiere teaching and research hospice. St. Christopher's is a well-designed, purpose-built, 62-bed unit located in Sydenham, a London suburb. Her previous patient, the Polish refugee, had left five hundred pounds (about $1000) as a legacy for the building.

No other hospice operates exactly like St. Christopher's. Each of the hundreds of hospices in Britain, the United States, and Canada has developed differently, according to patient needs, funding, staff personalities, and other factors. However, despite the many forms that hospice may take, there are some unifying principles and common aims.

Let's outline the standard oncology approach to see why the philosophy and techniques of hospice medicine or palliative care are needed to complement traditional medicine's attempts to cure at all costs.

TREATMENT OF CANCER AND AIDS PATIENTS IN AMERICA

I have spent all of my career in teaching hospitals, and so will now try to describe the experience of a cancer or AIDS patient in one of these facilities. Cancer patients being treated at teaching hospital-associated cancer clinics generally consult with medical residents, oncology fellows (trainees in the oncology specialty), and oncology faculty members (professors). These patients receive conventional anticancer therapies, including surgery, chemotherapy, and radiotherapy, or experimental treatments. If the treatments are not curative for the cancer, the patients may nonetheless hope for a temporary remission of cancer-related symptoms and for prolongation of life.

Intermittently, patients may need to be admitted to the hospital for some complicated forms of cancer therapy, such as chemotherapy infusions, or the initiation of a DHPG infusion for an eye infection in AIDS patients. Treatment of serious complications of the cancer as well as the cancer treatment itself also requires hospitalization. When in the hospital, the patient is treated by a full spectrum of residents, interns, and students, all under faculty supervision.

On admission to the hospital, the patient undergoes a routine battery of laboratory examinations, including blood tests, urinalysis, chest X-ray, and electrocardiogram. The medical students, interns, and residents question and examine the patient, usually independently. This may take three hours or more and can be exhausting for someone weakened by chronic illness.

One or two days after admission, the patient may begin to see specialist consultants requested by the intern, resident, or attending physician. A team of such consultants may consist of a student, resident, and faculty member with special expertise in cancers of the type to be treated. More questioning and examining is generally required, and this may lead to more sophisticated diagnostic tests.

Repeat blood tests ordered by the primary care doctors and specialists are usually continued every day. Running out of patient venipuncture and IV sites often becomes a major problem. And the pain from these procedures takes its toll on the bravest of patients.

Specialized tests such as X-rays or nuclear medicine scans may take several days to complete. Because of frequent procedures and interviews, the patient may miss meals or the food may be cold. Even at its best, hospital cooking cannot compare to home meals. Microwave ovens may be used to reheat a meal that had to be missed. After a while, most patients stop complaining about the food and resign themselves to accepting it.

House staff and faculty oncologists focus on the specific surgery, radiation, or chemotherapy treatment options for the particular cancer involved. Treatment of symptoms of the cancer—such as pain, sore mouth, vomiting, insomnia, or constipation—falls to the intern, or the junior member of the team. Teaching conferences usually deal with diagnosis and treatment of the cancer or its complications. They are less frequently concerned with difficult problems in the management of symptoms related to the cancer or with the psychological or social complications of the disease.

In order to be certain that the young physicians-in-training learn about as many types of cancer as possible, doctors' ward assignments are changed as frequently as every month. Thus, you may see ten or more doctors if you are hospitalized for several weeks. Each morning the patient may be tended by a medi-

* * * * *

An often-quoted study found that nurses take...twice as long, on average, to answer the calls of dying patients compared to other patients. Is it possible that doctors and nurses see dying or incurable patients as medical failures to be forgotten?

* * * * *

cal team consisting of five or more people in white coats. Frequently, patients cannot recall who their primary doctor may be. All of this makes it especially difficult to bring up sensitive matters and discuss fears and concerns.

Oncologists practicing privately in community hospitals keep equally busy and may also concentrate primarily on questions of diagnosis and anticancer therapy. For them too, pain and other cancer symptoms may or may not be optimally managed for the patient's comfort. Private oncologists, after all, were trained in academic teaching centers.

Hospital nurses are extremely busy and may, like other health care providers, be uncomfortable caring for the dying. So they may spend more time with those patients who are going to recover. They must take vital signs, check intravenous feeding lines, give medications, and perform numerous other routine tasks. The pressures of their duties make it difficult to give the extra attention that dying cancer and AIDS patients need. An often-quoted study found that nurses take about twice as long, on average, to answer the calls of dying patients compared to other patients. Is it possible that doctors and nurses see dying or incurable patients as medical failures to be forgotten? In short, the care afforded such patients is often less deeply committed than that given likely survivors.

Uncomfortable and expensive diagnostic tests, along with toxic anticancer treatments, are often used in last-ditch efforts to prolong life. I have seen cancer patients receive radiation therapy, chemotherapy, kidney dialysis, and total parenteral nutrition (intravenous feeding) up to the day of death—high-tech medical care scarcely justified by the circumstances. And even as this goes on amid the hustle and bustle of hospital routine, the complex nonmedical needs of dying cancer and AIDS patients just as often receive little attention.

The questions and problems of family members place further pressures on busy doctors and nurses. Restricted visiting hours suit the convenience of the staff, who must work efficiently. A spouse or close friend can't ordinarily stay overnight to be near their loved one. Some hospitals forbid children to visit at all. For hygienic reasons, pets are never allowed in hospitals. In short, the dying patient often suffers in a special purgatorio of medicine's own creation.

Since malignancies more commonly develop in older people, many cancer patients depend on an elderly spouse or other relative for home care. This exhausting work taxes the most altruistic and devoted person's ability to cope. The spouse or relative may feel guilty if he or she becomes irritable or impatient with the dying loved one, and the relationship may be seriously damaged under this strain. Often a week or two in the hospital for "social reasons" can help the family situation greatly. However, in expensive, acute-care hospitals, cancer patients cannot be admitted merely for the convenience of the family, but only to treat substantial medical problems.

And finally, since medical care is so compartmentalized, oncology doctors and nurses rarely attend to the problems of grieving relatives after the death of the cancer patient. Referral to a counselor or support group may be proposed only after the occurrence of a serious depression or a suicide attempt.

Standard oncology treatments concentrate on healing the body. Although oncologists care about and consult with the patient's family, they focus most of their attention on the medical treatment of the patient. This generally includes skill and experience in diagnosing cancer-related problems, and administering anticancer therapies such as surgery, radiation, and chemotherapy. Oncologists work almost exclusively in hospitals and outpatient clinics with countless other skilled professionals. They rarely, if ever, make home visits or see patients in nursing homes. Oncologists cure about 40% of their cancer patients and extend the lives of many others. Most cures come from the surgical excision of a localized cancer, or its treatment with radiation therapy. Chemotherapy for cancers that have spread actually cures fewer than five percent of patients.

For patients with advanced cancer, the diagnostic skills and anticancer treatment expertise of oncologists may no longer be relevant. Indeed, for many patients, the overuse of the diagnostic and therapeutic technology may only pointlessly increase the pain and suffering of dying. And so the hospitals designed to cure people of disease have become especially frightening and unsuitable places for those with advanced and incurable cancers.

HOSPICE PHILOSOPHY AND SERVICES

The hospice philosophy encompasses more comprehensive care than standard medical treatment. It includes physical, psychological, social, and spiritual therapies. Physicians, nurses, counselors, clergy, social workers, occupational and physical therapists, and volunteers work as a team to provide various hospice services. All work together to help and comfort the cancer or AIDS patient and his or her family during the dying process, concentrating on symptom control and psychosocial care.

The patient's family is important to the hospice care team. The emotional support provided by the team serves to help family and friends cope with the crisis and grow emotionally from the experience. Hospice workers may even visit the family while the patient is in the hospital or an inpatient hospice facility.

Since most people would rather spend their last days at home, hospice attempts to make this possible by providing emotional and psychological support and expert medical services in the home. If symptoms run out of control, the patient may be admitted to an inpatient hospice bed. If the family needs a respite from the constant demands of caring for their loved one, the patient may be admitted to a nursing home for few days of "respite care."

Wherever the patient resides, the hospice team is available to help with problems 24 hours a day, seven days a week. The constant availability of the hospice team gives patients and families a feeling of security. They know that hospice professionals can help them in virtually any situation without the need for hospitalization.

Hospice workers believe that dying is a normal process. They allow that process to run its course without attempting to hasten or delay it. They act on the basis of the dictum: "Where cure is not possible, care is still needed." The hospice's clear focus on improving the quality of life by relieving pain and other distressing symptoms helps patients open up, helps them with the nurturing and healing of their significant personal relationships, and helps them to an often surprising spiritual growth.

Traditionally, hospice teams provide services on the basis of need. In England, medical insurance or other means of payment are not required for entry into a hospice program. English hospices' funding comes from voluntary contributions, grants from small foundations, and patient legacies, along with support from the National Health System.

* * * * *

*The hospice patient never hears that nothing
else can be done....No one's pain and
suffering are shrugged off with the old adage
"What cannot be cured must be endured."*

* * * * *

PHYSICAL, PSYCHOLOGICAL, SOCIAL, AND SPIRITUAL ASPECTS OF HOSPICE

The hospice patient never hears that nothing else can be done to control symptoms. No one's pain and suffering are shrugged off with the old adage "What cannot be cured must be endured." Though cure of the cancer or a prolongation of survival may not be possible, constant efforts are made to ease the tormenting symptoms. With the physical and mental distress relieved, patients can live fully until they die.

Besides pain, many other symptoms related to the cancer, AIDS, or related therapies must be skillfully treated. Opioid side effects such as constipation and dry mouth require laxatives and lozenges. Problems such as breathlessness, nausea, vomiting, poor appetite, or infections call for a medical staff that is experienced, skillful, and patient.

Extra nursing and counseling staff attend to the patient's physical and psychological needs. Hospice physicians keep diagnostic tests to a minimum and order only those tests that will directly lead to better symptom control. Doctors budget more time for visiting patients and come alone or with only one or two others, rather than the mainstream hospitals' large group of doctors and nurses that only intimidate patients and prevent them from discussing their fears and concerns.

As their disease progresses and patients face increasing uncertainty, they want to hear one thing: "No matter what happens, I am going to stand by you and help you in any way that I can." But the doctor–patient relationship, founded on trust, fostered by honesty, is poisoned by deceit. The patient hears the truth about each situation as it arises. But truth has a broad spectrum, with gentleness at one end and harshness at the other. Patients always prefer the gentle truth.

The hospice team endeavors to maintain their patients' self-respect by involving them in the decisions that concern them. Patients appreciate retaining as much independence and control as possible. Hospice staff also remember that, though the dying patients receive a great deal of attention and support, they often provide emotional support and comfort to those who will survive them.

Not only are the physical symptoms of cancer and AIDS treated, but their mental and psychological problems receive close attention. Although the majority of cancer patients remain alert and aware throughout most of the illness, cancers sometimes cause confusion or clouding of the consciousness by spreading to the brain or by causing metabolic imbalances.

Patients may fear insanity, which rarely occurs as a direct result of the cancer itself. There may be frustration because of memory impairment, or difficulties with concentrating may arise. Relatives are often distressed if the patient appears confused or undergoes changes in personality.

Adjusting the medications may help a great deal, as does the support and reassurance of the staff, which often stays long hours with the patient. Besides appropriate medical treatment, the confused patient needs a well-lit, quiet room with some familiar personal objects. Tranquilizers may also help ease the patient's distress in this situation.

The physical surroundings of an inpatient hospice scarcely resemble a hospital. Families bring in items from home to decorate the room. Special efforts are made to provide decent cuisine and to serve it hot. Hospice cooks usually cater to individual preferences, and wine or beer may be served.

Hospice staff schedule the daily activities to suit the patient's convenience and allow flexibility for individual needs. The staff encourages visits from friends and relatives. Visiting hours are flexible, and hospices provide for overnight accommodations for family members of inpatients. Children and pets may visit without restriction. The hospice team allows family members to aid in the daily care of the patient. This lessens the patients' feelings of helplessness and strengthens the bonds with their loved ones.

Family members may also need support, for instance, if they feel guilty that they can no longer take care of their loved ones. They may need to talk about ambivalent feelings or unresolved conflicts with the dying or dead relative. "If only I had been a better husband." "I should have made him go to the doctor sooner."

During the dying process, people are capable of much growth. They take care of unfinished business, resolve personal disputes, and perhaps take a step in a spiritual direction. These can all add to their peace of mind and help to comfort the surviving relatives and friends. Family members and close friends may also grow emotionally and spiritually through the experience of helping to care for a dying person who has expert hospice care. This may help the adjustment of the survivors, lessening their suffering from the physical or mental problems of the grieving process.

After the death of a spouse or close relative, the bereaved generally need to talk about it with understanding listeners. Unfortunately, their close friends may cease to visit because of their own anxieties and feelings of helplessness. The hospice team continues to provide counseling and emotional support for the bereaved family and friends to help them avoid the serious mental and physical complications that frequently accompany bereavement. Once the survivors recover from the grieving process, they often become hospice volunteers in order to return their debt of gratitude.

Ministers, priests, and other religious counselors will visit hospice patients if requested to do so. Some hospices have regular nonsectarian religious services, such as communion. Although the majority of hospice workers are strongly committed Christians, they do not proselytize while caring for their patients, recognizing that it would be improper to try to convert patients to any religion by taking advantage of fears and dependencies.

Patients can prepare for death whether or not they have a belief in God or an afterlife. Hospice patients frequently choose to focus attention on the spiritual aspects of their lives. This often helps them find peace and acceptance.

THE MEDICARE AND MEDICAID HOSPICE BENEFIT

In 1983 Congress passed legislation establishing the Medicare Hospice Benefit. This allows a hospice team, usually administered by a visiting nurse associate, to manage a terminally ill person's total care. In 1987 they extended this benefit to indigent patients receiving MediCaid. Unfortunately, because of a lack of awareness about hospice, as well as severe financial constraints, the current availability and quality of services is relatively limited. Here is how the benefit works:

The patient's doctor, or more commonly a nurse, explains the hospice approach and recommends that the patient sign up for the program. A local visiting nurses' association sends nurses, social workers, home health aides, and other professionals to the home to help the patient and his or her family. Hospice staff members monitor patients medically, instruct them on the correct use of medications, help them find social services, and provide emotional support for both patients and caregivers.

If their physical symptoms become uncontrolled, patients may be admitted to an acute inpatient facility for intensive treatment. Unfortunately, hospital beds are generally contracted for this purpose, since too few patients receive hospice care to support many inpatient hospices. Usually, pain or other symptoms can be controlled in a few days.

If the family and other caregivers become exhausted by the task of caring for a hospice patient, the patient may then be admitted for a few days to a nursing home for respite care. Wards dedicated solely to respite care for hospice patients are virtually nonexistent. So hospice program directors generally contract with nursing homes for these beds on an as-needed basis. At least this gives the family a modest break from the strain of providing around-the-clock care. The rest usually restores their physical and emotional energies so that they can return to caring for their loved one.

When hospice patients are close to death, nurses may be assigned around the clock to provide the skilled nursing care often needed to support the family in keeping their loved ones at home. The majority of hospice patients die at home, never requiring hospitalization, nursing home admission, or around-the-clock nursing.

CHAPTER 10

Integrating Life-Prolonging Therapy with Hospice

Treatment for the pain and suffering of the terminally ill today is intolerably poor. I don't blame euthanasia proponents for proposing a remedy. But it's the wrong remedy!

Based on my experience with thousands of dying cancer and AIDS patients, I have reached a position quite different from that of both euthanasia proponents and advocates of the status quo. We need not choose between either continuing the current suffering of the terminally ill or legalizing active euthanasia. The hospice approach provides a workable and far more humane alternative to both.

We do, however, need to transcend institutional and financial barriers to make excellent hospice care more widely available in America. To better understand those barriers, let us look at the structural differences between health care systems in England and here.

THE BRITISH MEDICAL SYSTEM

The British National Health Service began in 1948 to provide a tax-funded, centralized, comprehensive medical care system from the cradle to the grave. A limited number of specialists treat patients in larger referral centers. General practitioners manage the nonspecialty medical care, including that for patients with advanced cancer.

The British system of government-controlled, socialized medicine costs much less per capita than the American system. However, a chronic shortage of hospital beds in Britain causes annoying delays for nonemergency surgeries and other services. Both professional and nonprofessional hospital workers are underpaid and consequently may exhibit low morale. Strikes occur occasionally because of poor pay and poor conditions. Moreover, since few young oncologists and other specialists are given hospital appointments to practice, many emigrate, causing a brain drain of the trained and committed.

In Britain everyone is assigned a general practitioner (GP). To consult a specialist such as a surgeon or oncologist, patients must be referred by their GP. Oncologists treat cancer patients at one of about a dozen speciality hospitals.

Patients referred to British cancer hospitals receive first-rate care, but experience a far more conservative use of modern cancer diagnostic and treatment technology. A team of consultant oncologists and registrars (residents) in training decide on the course of treatment and deliver the therapy. If no useful anticancer treatment remains for a patient with an incurable form of cancer, the patient returns to the GP for terminal care. The GP then generally refers the patient to the local hospice.

The average British GP with 2000–3000 patients on his or her roster takes care of four or five dying cancer patients per year. Usually the GPs lack the time and expertise to devote to this demanding task. For these reasons, British GPs enthusiastically welcomed the hospice movement, recognizing that its implementation would help them and their patients.

In England, excellent hospice services are now widely available and relatively better reimbursed. Physicians may now devote their careers to quality palliative care and gain recognition as specialists.

ORGANIZATION OF CANCER CARE IN AMERICA

In America, private insurance, Medicare, Medi-Caid, and veteran's benefits pay for medical care. A significant number of people, perhaps 15%, remain uninsured and many more are underinsured. The medical care delivery system consists of an assortment of competing institutions, including preferred provider organizations, health maintenance organizations (Kaiser Permanente and the like), and university, community, veteran's, and county hospitals. Almost 25% of medical costs go to administration, versus less than 3% in the UK's socialized medicine.

American medical schools emphasize technology-intensive specialty and subspecialty training. Consequently, surgeons, radiotherapists, and medical oncologists are abundant, whereas family practitioners are scarce. Few training programs for hospice physicians exist in America.

* * * * *

In America, an excess of hospital beds, an oversupply and maldistribution of specialists, and an uncontrolled proliferation of expensive medical technologies contribute to a skyrocketing inflation of medical costs.

* * * * *

Hospitals compete by acquiring the latest CT scanners, MRIs, radiation therapy machines, and catheterization labs. They also vie for medical specialists trained in the use of the latest technologies. Duplication of services and overuse of expensive technology is inevitable. In fact, owing to economic pressures, the survival of health care providers often depends on it.

In America, an excess of hospital beds, an oversupply and maldistribution of specialists, and an uncontrolled proliferation of expensive medical technologies contribute to a skyrocketing inflation of medical costs. Government regulation has not yet worked to control escalating hospital bills.

The American cancer patient will probably be treated throughout the course of the illness by oncology specialists in surgery, radiotherapy, and chemotherapy (medical oncology). In many large cities, a surplus of oncologists leads to competition for patients. Anticancer therapy, such as radiation and chemotherapy, may continue along with intensive supportive treatment (antibiotics, intravenous nutrition) into the terminal phase of the illness. All of these therapies are associated with high costs, and do little or nothing to relieve pain and suffering.

THE HIGH COST OF TERMINAL CARE

One-half of all cancer deaths in the United States occur among those age 65 and over. Medicare provides medical insurance for most Americans 65 and over. Gerald Riley and his colleagues analyzed medicare costs, broken down by cause of death, in 1979 (the last year for which data are available).[1] I will summarize their analysis as it relates to the inadequate delivery of palliative care services in America.

About one-fourth of Medicare expenditures occur in the last year of life. When Medicare expenditures are broken down by cause of death, the last year of life for cases of cancer costs twice as much as cases of heart disease ($8021 vs $4012). Hospitalization charges account for 76% of medical expenditures and physician and other medical fees consume 18%.

Average Medicare costs for cancer patients over 85 years old ($5670) are less than those from 75–84 year olds ($7873), which are exceeded by costs for patients in the 65–74-year-old group ($8835). Although no data exist to determine the average medical cost per cancer death for patients under 65 years old, the above trend suggests that they will be much higher than in the Medicare population.

For Medicare patients, cancer care expenditures are heavily concentrated in the last year of life. Thirty-four percent of charges occur in the last month of life and 59% in the last three months. The National Center for Health Statistics reported that in 1980, among 65–74-year-old people dying of cancer, 59.9% died in hospital.[2] This helps explain the high cost in the last month of life.

May Baker and her colleagues studied the medical costs of breast cancer patients, broken down by the phases of the illness. Hospitalization for initial treatment cost only 38% as much as the hospital costs of terminal care. Physicians receive 2.7 times the reimbursement for terminal care as for initial, potentially curative, treatment.[3]

Most cancer patients, including those with breast cancer, do not want hospitalization in the terminal phase of an incurable malignant disease. Hospitalization is rarely required when patients receive good palliative care. The high cost of terminal care indicates that palliative care services for breast cancer patients are underutilized.

Since 1979 medical costs have risen approximately 400%. Also since 1979, cancer deaths have gone up about 0.4% per year (now about 22% of deaths), whereas cardiovascular mortality has fallen about 1% per year (now about 40% of deaths). In the United States in 1991, 514,000 people died of cancer. Adjusting for inflation, the medical costs for the last year of life of these people was $18.0 billion, or 51% of the $35.3 billion spent for treating cancer.

Fred Hellinger, director of the division of cost and financing at the federal Agency for Health Care Policy and Research, reported that the cost of treating

AIDS and HIV-infected people reached $5.8 billion in 1991. He predicted that this cost would rise to $10.4 billion in 1994. Hospital fees constitute 75% of the total AIDS and HIV bill.[4]

In contrast to the astronomical cost of conventional cancer and AIDS treatment, hospice services receive a small fraction of reimbursements. In 1991 $479 million went for services to certified Medicare hospice benefit patients. Overall reimbursement for hospice services accounted for less than 4% of the yearly costs for medical treatment in the last year of life of cancer patients. Based on Los Angeles County Department of Health Services data, hospice services to AIDS patients amounted to less than 1% of the AIDS treatment costs in 1990.

FINANCIAL DISINCENTIVES TO GOOD PALLIATIVE CARE

During a recent lecture on cancer pain management, I polled a group of medical students, interns, and residents on their primary motivations for becoming physicians. A substantial number indicated that they wanted to save or prolong lives. However, the majority of those responding reported that their primary motivation was to relieve pain and suffering. No one admitted that financial reward was a primary factor underlying their personal decisions to become physicians.

Despite the noble aspirations of those entering medicine and many of those already in the medical establishment, we have a "money driven" rather than an "ideal driven" bureaucracy administering our health care delivery and medical education systems. Well-reimbursed services such as hospitalization are much more available compared to such poorly reimbursed services as hospice.

Inpatient hospitalization for terminal care is well reimbursed by Medicare, Medicaid, and insurance companies. Each hospital day required to treat patients with cancer or AIDS pays the hospital $1000 or more. Hospitals depend heavily on the terminally ill population for their economic survival.

Patients with advanced cancer and AIDS go to hospitals as a last resort when pain or other symptoms cannot be controlled at home. However, to justify acute care days in hospital, patients *must* receive active treatment intended to prolong life. These treatments may include intravenous antibiotics, hyperalimentation, chemotherapy, radiation therapy, or surgery. Terminally ill patients may also receive such sophisticated tests as MRIs and CT, bone, and ultrasound scans. Both treatments and tests are expensive.

For most hospitals and physicians, hospice care decreases income, and so loses money. If a hospice program is successful, most cancer and AIDS patients spend their last weeks and months at home, where they prefer to be, and costs are much lower. Hospice patients also receive many fewer high technology tests, and once again overall costs are lower.

Therefore, repeated admissions to hospitals for control of pain and other symptoms during the ter-

* * * * *

*...repeated admissions to hospitals
for control of pain and other symptoms
during the terminal phase is to the financial
advantage of hospitals and physicians.
This leads to a paradoxical situation in which
it pays to treat pain and other symptoms
of terminal disease inadequately.*

* * * * *

minal phase is to the financial advantage of hospitals and physicians. This leads to a paradoxical situation in which it pays to treat pain and other symptoms of terminal disease inadequately. The better job a doctor or hospital does in pain and symptom management, the more likely patients are to remain at home until they die of their disease. These well-managed patients simply require fewer hospital days and high-tech medical services, and the hospital and physicians lose potential revenue.

The adminstration of hospice benefits by federal bureaucrats and insurance companies also adds to the financial disincentives for hospice care in America. The scheduled Medicare and MediCaid hospice benefits set unrealistically low limits on reimbursement for patient services. No matter how many hospice services are required, the maximum payment is about $10,000 (1991 dollars), a small fraction of the cost of acute care services.

However, hospice services are also limited by a fixed per diem reimbursement for each patient. This reimbursement amounts on average to $3000 per month (1991 dollars) versus over $12,000 for the last month of life for patients remaining in the conventional care system. These budgetary contraints on certified Medicare and Medicaid hospice providers force them to reject many otherwise eligible patients who require palliative radiation therapy, IVs, or other expensive treatments.

The financial pressure in our system also leads to late referrals of terminally ill patients to hospice programs. The average length of stay, from admission to death, in the visiting nurse association-affiliated hospice where I currently work is about 30 days. Terminally ill patients in England average much longer in their hospice programs.

Although no one intended it to happen, our medical care system has evolved in such a fashion that good pain and symptom management for the terminally ill are financially discouraged. If a hospital delivers superb palliative care to terminally ill patients, it will take in far less revenue than its competitors and thus will soon go out of business.

Good palliative care receives little more than lip service from the American medical establishment. Physicians concentrating on palliative care in the United States are not recognized as specialists and earn far less than oncologists or other medical specialists. So American physicians not only suffer inadequate training in pain and symptom management, but also confront financial disincentives to the practice of good palliative care.

A PALLIATIVE CARE UNIT PROPOSAL

To change the unfortunate state of palliative care in America, we need a concerted effort by health care professionals, medical administrators, politicians, medical insurance agencies, and the public. The obstacles are so many that sometimes it may seem hard to know where to start. Good palliative care will not come about solely from the initiatives of politicians or medical administrators. It will require many individual demonstration projects all over the country, and clear evidence of their success.

I am currently making a specific proposal for a palliative care unit to the Los Angeles County Department of Health Services (DHS) that will constitute a step in the right direction. Let me sketch the background and specific recommendations.

Los Angeles County has three large university-affiliated teaching hospitals that serve primarily indigent patients. These hospitals are LA County—USC Medical Center, where I work; Harbor-UCLA Medical Center; and the Martin Luther King Medical Center. The combined inpatient average daily census of AIDS patients in 1991 was 75. At least 75 cancer patients are also in these hospitals each day. At any given time, about one-third of these inpatients have far-advanced disease for which the burdens of aggressive antitumor or antivirus treatment outweigh the benefits.

For the past 12 years, I have proposed a variety of plans for an inpatient palliative care ward. LA County DHS administrators have always acknowledged the need for and potential value of this kind of ward, though they always respond that the budget has no money for palliative care services. Their analyses have shown that good palliative care might save money for our MediCal insurance program (California's version of Medicaid) and LA County taxpayers, but it would cost money for the LA County DHS. Since these medical administrators work for DHS—not MediCal or the LA taxpayer—they could not start a palliative care program.

Finally, federal funds for AIDS services distributed to Los Angeles by the Ryan White Act provided an opportunity to help fund a palliative care ward in one of the LA county hospitals. I drafted a detailed proposal to present to the Ryan White Act commissioners in Los Angeles. It showed that $400,000 of seed money from the Ryan White Act could help start a 15-bed palliative care ward and associated consultation services at the three large county hospitals. Once initiated, this ward and the associated consultation services could subsequently survive economically.

In my research I contacted administrators from existing palliative care wards in North America, including Montreal, Quebec; Vancouver, British Columbia; San Francisco; and the Kaiser Hospitals in Los Angeles. Costs to provide acute-level palliative care services averaged about $600 per patient per day.

Darryl Nixon, chief MediCal administrator for a region of Los Angeles, agreed on the need for a palliative care ward, and offered to consider paying the regular acute-care-inpatient day fee (about $1000/day) for palliative care ward patients. In his analysis, this would save MediCal money by reducing the overall number of acute hospitalization days for terminally ill people.

This would constitute a win, win, win situation: Terminally ill patients would be given appropriate palliative care rather than attempts to prolong their dying with inappropriate and expensive high-tech care. The LA county hospital that housed the ward could make a profit serving the patients, and MediCal would save money overall from a reduced utilization of hospitalization services. Medical and nursing students, oncology fellows, and medical residents, could also receive palliative care training on this ward and the associated consultation services.

The LA County DHS administrators committee determining allocation of the Ryan White Act funds declined to provide seed money for the proposed palliative care ward. However, since my research has also suggested that the ward would do well economically without the Ryan White seed money, I pursued the proposal with DHS administrators.

After nearly a year of prodding, the director of the DHS asked administrators to meet with me about this proposal. As I write, officials of the Rancho Los Amigos Medical Center are analyzing my proposal before deciding whether to meet with MediCal.

The campaign to legalize euthanasia in California will help focus attention on positive alternatives, such as this one.

HOSPICE—ALTERNATIVE TO EUTHANASIA

Ideally the caring and life-prolonging techniques of oncology treatment and the caring and supportive approach of hospices could work together throughout the illness.

Rather than legalizing euthanasia, we need to promote, expand, and improve the hospice approach. We need to train more physicians, nurses, and other health care workers in hospice medicine. Rather than designating isolated beds in acute care hospitals for hospice care, we should have more inpatient palliative care wards available with experienced medical, nursing, and allied health care staff to manage difficult cases. The inpatient palliative care facilities could also serve as training sites for physicians and nurses in hospice medicine.

We need to introduce the hospice approach into the required education of our health care professionals. The reallocation of our cancer and AIDS treatment resources to increase the proportion going to hospice care is essential. Mechanisms must be developed to integrate pain control and palliative care into cancer and AIDS treatment from the time of the diagnosis. The public—and particularly cancer patients—need to be educated about hospice care.

We do not necessarily need British-style socialized medicine to improve the availability and qual-

* * * * *

With widespread availability of good hospice care for the terminally ill, the question of euthanasia would become moot.

* * * * *

ity of palliative care in this country. However, true health care reform in this country needs to address these perverse economic incentives, which are major barriers to the improvement of availability and the quality of hospice care. Adopting national health insurance similar to the system in Canada would help remove institutionalized financial barriers to improving hospice care.

The quality of life of the terminally ill can be improved with the expert attention to pain and other physical symptoms, as well the psychological, social, and spiritual support of the patient and his or her caregivers.

With widespread availability of good hospice care for the terminally ill, the question of euthanasia would become moot. People simply don't want euthanasia when they are physically comfortable and their emotional needs are addressed.

We have the knowledge and the means to assure that no terminally ill person need beg for death to end his or her suffering. We need the resolve to spread this precious knowledge. Universally available, excellent quality hospice medicine is the life-affirming alternative to the hopelessness of euthanasia.

THE PROPOSED CALIFORNIA DEATH WITH DIGNITY ACT—

"Terminal Illness, Assistance with Dying"

California Civil Code, Title 10.5

Sec. 1. Title 10.5 (commencing with Section 2525) is added to Division 3 of part 4 of the Civil Code, to read:

2525. *Title:* This title shall be known and may be cited as the Death With Dignity Act.

2525.1. *Declaration of Purpose.* The people of California do declare:

Current state laws do not adequately protect the rights of terminally ill patients. The purpose of this Act is to provide mentally competent terminally ill adults the legal right to voluntarily request and receive physician aid-in-dying. This Act protects physicians who voluntarily comply with the request and provides strong safeguards against abuse. The Act requires the signing of a witnessed revocable Directive in advance and then requires a terminally ill patient to communicate his or her request directly to the treating physician.

Self-determination is the most basic of freedoms. The right to choose to eliminate pain and suffering, and to die with dignity at the time and place of our own choosing when we are terminally ill is an integral part of our right to control our own destinies. That right is hereby established in law, but limited to ensure that the rights of others are not affected. The right should include the ability to make a conscious and informed choice to enlist the assistance of the medical profession in making death as painless, humane, and dignified as possible.

Modern medical technology has made possible the artificial prolongation of human life beyond natural limits. This prolongation of life for persons with terminal conditions may cause loss of patient dignity and unnecessary pain and suffering, for both the patient and the family, while providing nothing medically necessary or beneficial to the patient.

In recognition of the dignity which patients have a right to expect, the State of California recognizes the right of mentally competent terminally ill adults to make a voluntary revocable written Directive instructing their physician to administer aid-in-dying to end their life in a painless, humane and dignified manner.

The Act is voluntary. Accordingly, no one shall be required to take advantage of this legal right or to participate if they are religiously, morally or ethically opposed.

2525.2 *Definitions.* The following definitions shall govern the construction of this title:

(a) "Attending physician" means the physician selected by, or assigned to, the patient who has primary responsibility for the treatment and care of the patient.

(b) "Directive" means a revocable written document voluntarily executed by the declarant in accordance with the requirements of Section 2525.3 in substantially the form set forth in Section 2525.24.

(c) "Declarant" means a person who executes a Directive, in accordance with this title.

(d) "Life-sustaining procedure" means any medical procedure or intervention which utilizes mechanical or other artificial means to sustain, restore, or supplant a vital function, including nourishment and hydration which, when applied to a qualified patient, would serve only to prolong artificially the moment of death. "Life-sustaining procedure" shall not include the administration of medication or the performance of any medical procedure deemed necessary to alleviate pain or reverse any condition.

(e) "Physician" means a physician and surgeon licensed by the Medical Board of California.

(f) "Health care provider" and "Health care professional" mean a person or facility or employee of a health care facility licensed, certified, or otherwise authorized by the law of this state to administer health care in the ordinary course of business or practice of a profession.

(g) "Community care facility" means a community care facility as defined in Section 1502 of the Health and Safety code.

(h) "Qualified patient" means a mentally competent adult patient who has voluntarily executed a currently valid revocable Directive as defined in this section, who has been diagnosed and certified in writing by two physicians to be afflicted with a terminal condition, and who has expressed an enduring request for aid-in-dying. One of said physicians shall be the attending physician as defined in subsection (a). Both physicians shall have personally examined the patient.

(i) "Enduring request" means a request for aid-in-dying, expressed on more than one occasion.

(j) "Terminal condition" means an incurable or irreversible condition which will, in the opinion of two certifying physicians exercising reasonable medical judgment, result in death within six months or less. One of said physicians shall be the attending physician as defined in subsection (a).

(k) "Aid-in-dying" means a medical procedure that will terminate the life of the qualified patient in a painless, humane and dignified manner whether administered by the physician at the patient's choice or direction or whether the physician provides means to the patient for self-administration.

2525.3 *Witnessed Directive.* A mentally competent adult individual may at any time voluntarily execute a revocable Directive governing the administration of aid-in-dying. The Directive shall be signed by the declarant and witnessed by two adults who at the time of witnessing, meet the following requirements:

(a) Are not related to the declarant by blood or marriage, or adoption;

(b) Are not entitled to any portion of the estate of the declarant upon his/her death under any will of the declarant or codicil thereto then existing, or, at the time of the Directive, by operation of law then existing;

(c) Have no creditor's claim against the declarant, or anticipate making such claim against any portion of the estate of the declarant upon his or her death.

(d) Are not the attending physician, an employee of the attending physician, a health care provider, or an employee of a health care provider;

(e) Are not the operator of a community care facility or an employee of a community care facility.

The Directive shall be substantially in the form contained in Section 2525.24.

2525.4 *Skilled Nursing Facilities:* A Directive shall have no force or effect if the declarant is a patient in a skilled nursing facility as defined in subdivision (c) of Section 1250 of the Health and Safety Code and intermediate care facility or community care facility at the time the Directive is executed unless one of the two witnesses to the Directive is a Patient Advocate or Ombudsman designated by the Department of Aging for this purpose pursuant to any other applicable provision of law. The Patient Advocate or Ombudsman shall have the same qualifications as a witness under Section 2525.3.

The intent of this paragraph is to recognize that some patients in skilled nursing facilities may be so insulated from a voluntary decision-making role, by virtue of the custodial nature of their care, as to require special assurance that they are capable of willingly and voluntarily executing a Directive.

2525.5 *Revocation:* A Directive may be revoked at any time by the declarant, without regard to his or her mental state or competency, by any of the following methods:

(a) By being canceled, defaced, obliterated, burned, torn, or otherwise destroyed by or at the direction of the declarant with the intent to revoke the Directive.

(b) By a written revocation of the declarant expressing his or her intent to revoke the Directive, signed and dated by the declarant. If the declarant is in a health care facility and under the care and management of a physician, the physician shall record in the patient's medical record the time and date when he or she received notification of the written revocation.

(c) By a verbal expression by the declarant of his or her intent to revoke the Directive. The revocation shall become effective only upon communication to the attending physician by the declarant. The attending physician shall confirm with the patient that he or she wishes to revoke, and shall record in the patient's medical record the time, date and place of the revocation.

There shall be no criminal, civil or administrative liability on the part of any health care provider for following a Directive that has been revoked unless that person has actual knowledge of the revocation.

2525.6 *Term of Directive:* A Directive shall be effective unless and until revoked in the manner prescribed in Section 2525.5. This title shall not prevent a declarant from re-executing a Directive at any time in accordance with Section 2525.3, including re-execution subsequent to a diagnosis of a terminal condition.

2525.7 *Administration of Aid-in-Dying:* When, and only when, a qualified patient determines that the time for physician aid-in-dying has arrived and has made an enduring request, the patient will communicate that determination directly to the attending physician who will administer aid-in-dying in accordance with this Act.

2525.8 *No Compulsion:* Nothing herein requires a physician to administer aid-in-dying, or a licensed health care professional, such as a nurse, to participate in administering aid-in-dying under the direction of a physician, if he or she is religiously, morally or ethically opposed. Neither shall privately owned hospitals be required to permit the administration of physician aid-in-dying in their facilities if they are religiously, morally or ethically opposed.

2525.9 *Protection of Health Care Professionals:* No physician, health care facility or employee of a health care facility who, acting in accordance with the requirements of this title, administers aid-in-dying to a qualified patient shall be subject to civil, criminal, or administrative liability therefore. No licensed health care professional, such as a nurse, acting under the direction of a physician, who participates in the administration of aid-in-dying to a qualified patient in accordance with this title shall be subject to any civil, criminal, or administrative liability. No physician, or licensed health care professional acting under the direction of a physician, who acts in accordance with the provisions of this chapter, shall be guilty of any criminal act or of unprofessional conduct because he or she administers aid-in-dying.

2525.10 *Transfer of Patient:* No physician, or health care professional or health care provider acting under the direction of a physician, shall be criminally, civilly, or administratively liable for failing to effectuate the Directive of the qualified patient, unless there is willful failure to transfer the patient to any physician, health care professional, or health care provider upon request of the patient.

2525.11 *Fees:* Fees, if any, for administering aid-in-dying shall be fair and reasonable.

2525.12 *Independent Physicians:* The certifying physicians shall not be partners or shareholders in the same medical practice.

2525.13 *Consultations:* An attending physician who is requested to give aid-in-dying may request a psychiatric or psychological consultation if that physician has any concern about the patient's competence, with the consent of a qualified patient.

2525.14 *Directive Compliance:* Prior to administering aid-in-dying to a qualified patient, the attending physician shall take reasonable steps to determine that the Directive has been signed and witnessed, and all steps are in accord with the desires of the patient, expressed in the Directive and in their personal discussions. Absent knowledge to the contrary, a physician or other health care provider may presume the Directive complies with this title and is valid.

2525.15 *Medical Standards:* No physician shall be required to take any action contrary to reasonable medical standards in administering aid-in-dying.

2525.16 *Not Suicide:* Requesting and receiving aid-in-dying by a qualified patient in accordance with this title shall not, for any purpose, constitute a suicide.

2525.17 *Insurance*: (a) No insurer doing business in California shall refuse to insure, cancel, refuse to renew, re-assess the risk of an insured, or raise premiums on the basis of whether or not the insured has considered or completed a Directive. No insurer may require or request the insured to disclose whether he or she has executed a Directive.

(b) The making of a Directive pursuant to Section 2525.3 shall not restrict, inhibit, or impair in any manner the sale, procurement, issuance or rates of any policy of life, health, or disability insurance, nor shall it affect in any way the terms of an existing policy of life, health or disability insurance. No policy of life, health, or disability insurance shall be legally impaired or invalidated in any manner by the administration of aid-in-dying to an insured qualified patient, notwithstanding any term of the policy to the contrary.

(c) No physician, health care facility, or other health care provider, and no health care service plan, insurer issuing disability insurance, other insurer, self-insured employee welfare benefit plan, or nonprofit hospital service plan shall require any person to execute or prohibit any person from executing a Directive as a condition for being insured for, or receiving, health care services, nor refuse service because of the execution, the existence, or the revocation of a Directive.

(d) A person who, or a corporation, or other business which, requires or prohibits the execution of a Directive as a condition for being insured for, or receiving, health care services is guilty of a misdemeanor.

(e) No life insurer doing business in California may refuse to pay sums due upon the death of the insured whose death was assisted in accordance with this Act.

2525.18 *Inducement:* No patient may be pressured to make a decision to see aid-in-dying because that patient is a financial, emotional or other burden to his or her family, other persons, or the state. A person who coerces, pressures or fraudulently induces another to execute a Directive under this chapter is guilty of a misdemeanor, or if death occurs as a result of said coercion, pressure or fraud, is guilty of a felony.

2525.19 *Tampering:* Any person who willfully conceals, cancels, defaces, obliterates, or damages the Directive of another without the declarant's consent shall be guilty of a misdemeanor. Any person who falsifies or forges the Directive of another, or willfully conceals or withholds personal knowledge of a revocation as provided in Section 2525.5, with the intent to induce aid-in-dying procedures contrary to the wishes of the declarant, and thereby, because of such act, directly causes aid-in-dying to be administered, shall be subject to prosecution for unlawful homicide as provided in Chapter 1 (commencing with Section 187) of Title 8 of Part 1 of the Penal Code.

2525.20 *Other Rights:* This Act shall not impair or supersede any right or legal responsibility which any person may have regarding the withholding or withdrawal of life-sustaining procedures in any lawful manner.

2525.21 *Reporting:* Hospitals and other health care providers who carry out the Directive of a qualified patient shall keep a record of the number of these cases, and report annually to the State Department of Health Services the patient's age, type of illness, and the date the Directive was carried out. In all cases, the identity of the patient shall be strictly confidential and shall not be reported.

2525.22 *Recording:* The Directive, or a copy of the Directive, shall be made a part of a patient's medical record in each institution involved in the patient's medical care.

2525.23 *Mercy Killing Disapproved:* Nothing in this Act shall be construed to condone, authorize, or approve mercy killing.

2525.24 *Form of Directive:* In order for a Directive to be valid under this title, the Directive shall be in substantially the following form:

Voluntary Directive to Physicians

> *Notice to Patient:* This document will exist until it is revoked by you. This document revokes any prior Directive to administer aid-in-dying, but does not revoke a durable power of attorney for health care or living will. You must follow the witnessing procedures described at the end of this form or the document will not be valid. You may wish to give your doctor a signed copy.

INSTRUCTIONS FOR PHYSICIANS

Administration of a Medical Procedure to End My Life in a Painless, Humane, and Dignified Manner

This Directive is made this ____ day of _____ (month) _____ (year).

I, _____, being of sound mind, do voluntarily make known my desire that my life shall be ended with the aid of a physician in a painless, humane, and dignified manner when I have a terminal condition or illness, certified to be terminal by two physicians, and they determine that my death will occur within six months or less.

When the terminal diagnosis is made and confirmed, and this Directive is in effect, I may then ask my attending physician for aid-in-dying. I trust and hope that he or she will comply. If he or she refuses to comply, which is his or her right, then I urge that he or she assist in locating a colleague who will comply.

Determining the time and place of my death shall be in my sole discretion. The manner of my death shall be determined jointly by my attending physician and myself.

This Directive shall remain valid until revoked by me. I may revoke this Directive at any time.

I recognize that a physician's judgment is not always certain, and that medical science continues to make progress in extending life, but in spite of these facts, I nevertheless wish aid-in-dying rather than letting my terminal condition take its natural course.

I will endeavor to inform my family of this Directive, and my intention to request the aid of my physician to help me to die when I am in a terminal condition, and take those opinions into consideration. But the final decision remains mine. I acknowledge that it is solely my responsibility to inform my family of my intentions.

I have given full consideration to and understand the full import of this Directive, and I am emotionally and mentally competent to make this Directive. I accept the moral and legal responsibility for receiving aid-in-dying.

This Directive will not be valid unless it is signed by two qualified witnesses who are present when you sign or acknowledge your signature. The witnesses must not be related to you by blood, marriage, or adoption; they must not be entitled to any part of your estate or at the time of execution of the Directive have no claim against any portion of your estate, nor anticipate making such claim against any portion of your estate; and they must not include: your attending physician; an employee of the attending physician; a health care provider; an employee of a health care provider; the operator of the community care facility or an employee of an operator of a community care facility.

If you have attached any additional pages to this form, you must sign and date each of the additional pages at the same time you date and sign this Directive.

Signed: _____

_____
City, County, and State of Residence

Proponents of Initiative

Robert L. Risley, JD
Chairman of the Board
Americans Against Human Suffering, Inc.
1320 So. Oak Knoll Ave.
Pasadena, CA 91106

Allan K. Briney, MD
Past President
Los Angeles County Medical Association
13303 E. Hadley Street
Whittier, CA 90601

Statement of Witnesses

I declare under penalty of perjury under the laws of California that the person who signed or acknowledged this document is personally known to me (or proven to me on the basis of satisfactory evidence) to be the declarant of this Directive; that he or she signed and acknowledged this Directive in my presence, that he or she appears to be of sound mind and under no duress, fraud, or undue influence; that I am not the attending physician, a health care provider, an employee of a health care provider, the operator of a community care facility, or an employee of an operator of a community care facility.

I further declare under penalty of perjury under the laws of California that I am not related to the declarant by blood, marriage, or adoption, and, to the best of my knowledge, I am not entitled to any part of the estate of the principal upon the death of the principal under a will now existing or by operation of law, and have no claim nor anticipate making a claim against any portion of the estate of the declarant upon his or her death.

Dated: _____
Witness's Signature: _____
Print Name: _____
Residence Address: _____

Dated: _____
Witness's Signature: _____
Print Name: _____
Residence Address: _____

Statement of Patient Advocate or Ombudsman

(If you are a patient in a skilled nursing facility, one of the witnesses must be a Patient Advocate or Ombudsman. The following statement is required only if you are a patient in a skilled nursing facility, a health care facility that provides the following basic services: skilled nursing care and supportive care to patients whose primary need is for availability of skilled nursing care on an extended basis. The Patient Advocate or Ombudsman must sign the "Statement of Witnesses" above AND must also sign the following statement.)

I further declare under penalty of perjury under the laws of California that I am a Patient Advocate or Ombudsman as designated by the State Department of Aging and that I am serving as a witness as required by Section 2525.4 of the California Civil Code.

Signed: _____

APPENDIX II

Americans Against Human Suffering Solicitation Letter from Richard Dreyfuss

Dear Reader,

When I did the research that went into my role in "Whose Life Is It, Anyway?" I was appalled to discover for the first time the true cost of this country's archaic and authoritarian medical practices.

In the name of preserving life, too often those practices condemn innocent men and women to weeks, months and years of hopeless, useless pain from which the sufferer prays for merciful release.

Sometimes those prayers are answered. More often, a friend or loved one, as we read in the papers increasingly, shoulders the responsibility of putting an end to the pain.

Such an act is called murder and can be punished as such. Or, if the sick person kills himself, his death carries the stigma of suicide.

I happen to believe that one's life belongs to oneself and that the individual—not any doctor, hospital or clergymen— has the right at least to refuse pointless treatment.

I hope you will join me in this struggle to change the archaic laws in our country. It may be the first step toward a more humane society for you and me and everyone.

Sincerely,

Richard Dreyfuss

COMMENTS

Dreyfuss' point about condemning innocent people to hopeless, useless pain gets back to the need to improve physician and nurse training in pain and symptom management for the terminally ill.

When we read in the newspaper about "mercy killing" by friends or loved ones, the person killed often is not terminally ill. The quadriplegic whose role Mr. Dreyfuss played was not terminally ill. Neither are senile or arthritic people. They would not be included under the proposed California Death with Dignity Act.

APPENDIX III

Americans Against Human Suffering Solicitation Letter from Ruth Bartling

Dear Mr. Cundiff,

It should never happen to you.

My husband, William Bartling, was near the end of his road when he entered the Adventist Medical Center of Glendale, CA, in April 1984.

He was 70. His health had been ebbing for six years. He was suffering from at least three illnesses—any one of which could kill him.

Then his doctors spotted a new shadow on his chest X-Rays: inoperable lung cancer.

When they inserted a biopsy needle into the tumor, his lung collapsed. It failed to reinflate, so they cut a hole through Bill's windpipe and used it to connect him to a respirator.

Then they gave up on him. "No medical treatment can result in any material improvement," one said in May 1984.

Still, they struggled to keep him alive.

But they could not help him die. That was against the law, and still is.

Bill Bartling lay for painful months, hopelessly alone in a tiny, windowless room amid life-support machines connected to tubes down his throat, up his nose and in his arms, shoulders and abdomen.

"I can't take it anymore," he told me again and again. When he tried to pull out his breathing tubes, the hospital staff strapped his hands to his bedframe.

Frustrated by hospital authorities, we hired a lawyer to ask court permission to remove the respirator so "the natural process of dying could occur peacefully, privately and with dignity."

Although my husband underlined his wishes through harrowing videotaped testimony to the court, the judge refused his request—not just once, but three times.

(Oddly, the court pointed out that the law in California, as in other states, permits life-support machines to be removed from irreversibly comatose patients, but not from conscious, aware sufferers like my husband.)

Our lawyer said, "They have sentenced him to death prolonged as long as medically possible." He appealed to a higher court.

Still strapped in his bed, William Bartling died of kidney failure in Glendale Adventist Medical Center at 2:42 PM on November 6, 1984.

His merciful release came after six months of hopeless, useless suffering—just 23 hours before his case was to be heard in the Court of Appeal. The Court sat anyway—and finally ruled in Bill's favor on December 27. Their decision was too late for Bill. Long before, he should have been allowed to refuse treatment when he wanted to say "No" to the pain that was eating him alive.

However, I am glad to think that the verdict in my husband's case may make a quicker, more humane death than his possible for other suffering Americans in the future.

That's why I'm writing you on behalf of a new nonprofit organization.

Americans Against Human Suffering intends to change existing law to make freedom of choice for the patient a reality.

Today, as Bill and I found to our sorrow, doctors and hospitals can ignore the wishes of their patients and enforce unwanted treatment even when that treatment cannot save lives.

Most of these doctors, and most hospitals, are terrified of liability lawsuits. They persist in hopeless and unwanted treatments because, although both inhumane and medically useless, they are the legally safe thing to do.

This situation has to change!....

It is more than a matter of life and death.

Today, in America, it is a crime to let a cat or a dog suffer needless pain. And cruel and unusual punishment cannot be inflicted on even the most hateful criminal.

Innocent men and women should be so lucky!

Think of William Bartling. Then think of your own loved ones, and of yourself. And take the first step toward a better America. Thank you.

Yours Truly,

Ruth Bartling

COMMENTS

Unfortunately, Mr. Bartling's doctors did not withdraw his life-support machines themselves without involving the court. To resolve any concerns, the doctors should have taken the issue to the ethics committee of the hospital, which should have recommended honoring Mr. Bartling's wish to stop life-support treatment. Again, existing law already covers Mr. Bartling's situation. Because of his terminal illness and the futility of further treatment to cure him or prolong his life, he had asked for withdrawal of life support. He did not ask for active euthanasia or assisted suicide. Given the facts of the case, I don't know why the lower court held that he had to remain on the respirator against his will. The appeals court made the right decision according to existing law.

The individual doctors, hospital administrators, and legal advisors involved in this case caused Mr. Bartling's unnecessary pain and suffering because of their lack of understanding of palliative care. I found no indication that a hospice consultant was ever called.

APPENDIX IV

Americans Against Human Suffering Solicitation Letter from Bob Risley, President

Dear Mr. Cundiff,

Twenty-one months ago, my wife Darlena died of cancer.

Darlena was a beautiful, sensitive woman. But during the last ten days of her life, her pain was extraordinary. Finally, she died in my arms.

If Darlena had asked, I was mentally prepared to help her escape her agony and suffering. But her request never came.

In my heart I know that helping her to die would have been the right thing to do if she had asked me. But I also know that if I had acted to end her pain and suffering, I would have been branded a criminal, and could have gone to jail.

That's why I'm writing you today, asking you to sign the enclosed letter to your Congressman, Dan Lungren.

It urges him to support our Humane and Dignified Death Act.

When passed by individual States, that law will give the terminally ill the legal right to decide for themselves whether life-supporting machines and procedures be withheld or withdrawn. Further, the terminally ill—and *only* the terminally ill—will have the right to request a physician's aid in a voluntary, humane and dignified death. Let me explain . . .

Darlena was only 41 years old when she died. What angered me most about her death was not so much the loss of a young, vibrant woman in the prime of her life. It was how Darlena was forced to suffer. Her pain, in spite of the drugs her doctors prescribed, was sometimes unbearable, even for a woman as strong as Darlena.

My wife loved life and fought to live. More than anything else, she wanted to control her life. After fighting back with all her strength, however, it became clear to both of us that only death would end her pain.

As long as the doctors kept her body working, she would suffer. We knew Darlena's remaining days would either be clouded by morphine or be spent clenching her fists to fight the pain.

Yet, our doctors could do no more than prescribe additional drugs, or they would take the chance of being branded as criminals and put in jail.

You see, even when a doctor knows that hope is gone—that to continue the struggle against a terminal disease means only continued pain and suffering—he must legally keep his patient alive on the machines regardless of what the patient may want.

And when a doctor does decide to help, it's usually done without the patient having any control over the decision. Under current laws, if a physician were to discuss with the patient the alternative of death with dignity and without pain, and then—on instructions of the patient implement this alternative—the doctor could be accused of murder. That's the law!

Whatever the doctor may decide, the patient too often has absolutely no control over the decision. Either the physician aids death without discussing it with the patient, or the patient is forced to endure and suffer.

That's why I formed Americans Against Human Suffering (AAHS) to promote the Humane and Dignified Death Act. There are other groups that ably promote "living will" legislation, but they do not address the real need for a doctor's assistance in dying upon request. AAHS takes up this challenge.

I don't want any of your loved ones to have to suffer the way my wife Darlena was forced to. She was denied the right to choose a humane and dignified death.

I wish I could have given her that right. But the law stopped me. And now I'm asking you to help me change those laws.

First, it's important that you understand what the Humane and Dignified Death Act will do.

Our Act applies only to the terminally ill. It gives them—and only them—the legal right to obtain a physician's aid in a voluntary, humane and dignified end to their suffering.

A terminally-ill person will have the legal right to decide whether life-sustaining machines and procedures be withheld or withdrawn. And, if needed, the terminally ill will be able to request a physician to administer aid in dying.

The decision to extend the normal life of a terminally-ill patient will be the decision of the terminally-ill individual and no one else.

I'm not saying that a terminally-ill patient should be denied the use of life-extending machines. In fact, they should have them as long as they want them. But it should be the patient's own decision and not the choice of a doctor, hospital administrator, insurance company bureaucrat or anyone else.

The Humane and Dignified Death Act will give you and me this right. It's too late for my wife. However...

By further legalizing what has become known as a "Living Will," you will be able to instruct your physician to withhold or withdraw life-sustaining procedures when they provide nothing medically beneficial to you and only prolong your suffering. And, in cases like my wife's, terminally-ill patients will be able to request a doctor's assistance in a painless and dignified death to end their suffering.

The decision will be the patient's. The terminally ill will be able to decide in advance how long to lie in a coma on a life-support machine or how much pain and suffering to endure.

It's not a pleasant subject. But I wish Darlena had had the choice the Humane and Dignified Death Act will give us.

As you probably realize, it's going to be a tough battle getting our Act sponsored and then passed. Although a recent national Roper survey reveals that 62% of the American people support the right for a humane and dignified death, Congress and our State Legislatures have refused to debate the issue.

That's where AAHS comes in. Its job is to mobilize the American public and urge our elected officials to grant the terminally ill the right to make their own decisions. Still, it's not going to be an easy fight. It's an emotional issue that most politicians would rather ignore. A few may even go so far as to accuse AAHS of trying to kill off helpless terminal patients.

Nothing could be further from the truth, and I hope you realize that.

I simply want you and your loved ones to have a choice my wife didn't have...

...A choice between painful, hopeless, day-to-day suffering or a life ending humanely and painlessly with dignity.

I'm counting on you to sign your letter to your US Congressman and to mail it to me today...Please don't delay. We are truly on a shoe-string budget and I fear that without your strong support, someone near your may be forced to suffer as my wife, Darlena, did.

Bob Risley, President
Americans Against Human Suffering

COMMENTS

It is remarkable that, despite Darlena's pain and suffering, she never asked for euthanasia. This reinforces my previous point about the infrequency of requests for euthanasia and assisted suicide despite the very poor state of palliative care in America.

Again, uncontrolled pain seems to be the dominant factor underlying the issue of active euthanasia. Since his wife was probably under the care of a competent cancer specialist, it might never have occurred to Mr. Risley that more could have been done medically to control the pain. Almost always we can do more that just prescribing more drugs to control the pain (*see* Chapter 5—"On Pain and Living"). Mr. Risley makes no mention of efforts to control pain, or consulting pain management or hospice specialists.

Mr. Risley again confuses the issue about life-support technology. His previously proposed Humane and Dignified Death Act and the current California Death with Dignity Act are not needed to give the terminally ill the legal right to decide for themselves whether life-supporting machines and procedures should be withheld or withdrawn. Terminally ill people and everyone else already have the right to refuse medical treatment. Doctors are not legally bound to keep patients alive on machines regardless of what the patient may want.

The physician need not choose between either hastening death, without discussing it with the patient, or forcing the patient to endure and suffer. The other alternative is to take training in hospice medicine and become skilled in controlling the pain and symptoms of terminal disease.

I imagine that, if doctors had offered Darlena the choice of euthanasia or continued uncontrolled pain and suffering, she might have chosen euthanasia. However, she also needed, and was not given the chance, to choose expert palliative care with specialists in hospice.

APPENDIX V

Dr. Timothy Quill

In March of 1991, Dr. Timothy E. Quill of Genesee Hospital in Rochester, NY published an article in the *New England Journal of Medicine*[1] describing the case of a terminally ill leukemic woman whom he indirectly assisted to commit suicide. Dr. Quill depicted "Diane" (her true identity was later found to be Diane Trumbull) as a middle-aged woman with vaginal cancer as a young woman and a long history of depression and alcoholism.

She had been sober for three years when cancer specialists diagnosed acute leukemia. The hematologist consultant told Diane that chemotherapy and bone marrow transplantation offered about 25% chance of cure. This treatment would entail hair loss, probable infectious complications, and prolonged stays in the hospital. She decided not to undergo chemotherapy and received symptomatic treatment with blood transfusions and antibiotics for infections under a home hospice care program.

Dr. Quill told her that he would order pain medications and other treatments to keep her comfortable. However, Diane had known people who had lingered with terminal diseases in what was called relative comfort. She wanted no part of it. She discussed with Dr. Quill and her family her wish to take her own life in the least painful way possible. Dr. Quill referred her to the Hemlock Society where she received a medication recipe for suicide.

Over several months Diane developed bone pain, weakness, fatigue, fevers, and increasing dependence. Dr. Quill reported that she had to choose between pain and sedation. In this setting of physical and emotional suffering, she overdosed on barbituates prescribed by Dr. Quill and died alone.

Advocates of euthanasia and assisted suicide offer this case as a perfect example of the logic of these methods for terminally ill people to avoid needless suffering.

However, many questions remain unanswered from Dr. Quill's report of Diane's case. What was her previous experience with relatives or friends with cancer who had lingering, painful deaths? What kind of pain did Diane have and what treatment was prescribed? How did she take the pain medicines and how did they work? What treatment was prescribed for her recurrent bouts of depression?

Good hospice care can alleviate much of the pain, other physical symptoms, and psychological distress of advanced cancer. Mistakes in prescribing medication for managing cancer pain are common. If Diane's pain was treated inadequately, changing the medication would be a better solution than suicide. Depression associated with advanced cancer may often be successfully treated with medication and psychotherapy.

These important unanswered factors may explain what drove Diane to commit suicide.

APPENDIX VI

Dr. Jack Kevorkian and His "Suicide Machine"

As a pathology resident in the 1950s, Dr. Jack Kevorkian advocated allowing death-row prison inmates to volunteer for experimental surgical procedures. After their operations, the condemned prisoners would simply not be revived from the anesthetic and allowed to die. The University of Michigan Pathology Department dismissed Dr. Kevorkian from his residency for continuing to carry on this quixotic crusade.

Later, in the 1960s and 1970s, Dr. Kevorkian modified his proposal to permit the condemned prisoners to donate organs for transplantation into others. He visited prisons and wrote letters to death row prisoners attempting to prove that significant numbers would be interested in donating their organs. Dr. Kevorkian lobbied politicians to introduce legislation to allow executed prisoners' organs to be donated to others. Few people took him seriously.

In the 1980s, with much more attention directed toward the issues of euthanasia and assisted suicide, Dr. Kevorkian invented his suicide machines, which he called "mercytrons." He began advocating the establishment of "obitariums" for those needing deliverance from suffering.

"Alzheimer's Disease" patient Janet Adkins' suicide using a machine devised by Dr. Kevorkian has received prominent media attention as a "right-to-die" case. This case needs to be put into perspective with respect to current efforts to legalize assisted suicide and euthanasia for the terminally ill.

By all accounts 54-year-old Adkins had no physical deterioration. She had beaten her son in tennis the week before her death. On the day before her suicide, she wrote a statement explaining her decision to die. This well-written document showed no signs of mental deterioration. Doctors apparently told Adkins that she would gradually lose her mental and physical abilities over five to ten years. She argued that, if she waited, she would not be able to choose to kill herself or be capable of carrying out suicide.

The autopsy of her brain showed an early stage of Alzheimer's disease with no significant shrinkage of the brain yet evident.

Many people who favor legalizing euthanasia and assisted suicide do not believe the Adkins case is a good example with which to promote their cause. The Hemlock Society and other pro-euthanasia groups have not campaigned for the legalization of euthanasia for Alzheimer's Disease patients.

Truly demented people are not competent to understand about euthanasia and to give their informed consent. The ballot initiatives in Washington state and California proposed the legalization of euthanasia only for terminally ill people with a prognosis of less than six months.

In October 1991, Dr. Kevorkian, now widely known as Dr. Death, struck again. Using more "mercytrons," he assisted the suicides of Marjorie Wantz and Sherry Miller. Wantz, age 53, suffered from pelvic inflammatory disease, which is treatable, not incurable or even life-shortenting. Miller had multiple sclerosis with major disabilities, but again was not terminally ill.

Dr. Kevorkian now faces murder charges in the deaths of these two women.

Dr. Kevorkian's book, *The Goodness of a Planned Death,* describes his efforts to solicit patients to whom he might offer his doctor-assisted suicide services. He advertized in local Michigan newspapers for terminally ill patients and received very few replies. He then obtained a list of medical oncologists and phoned cancer doctors directly, requesting access to their terminally ill patients who wanted suicide or euthanasia. In 1987 Dr. Kevorkian called me and asked whether I could refer him suitable candidates for assisted suicide.

He also wanted to meet with me to discuss my views on euthanasia and assisted suicide for cancer patients. He asked how many times I had performed euthanasia on my cancer patients. I told him that I had never assisted suicide or performed euthanasia and that, in my experience, good pain and symptom management make euthanasia unnecessary. Dr. Kevorkian expressed annoyance and impatience, saying that a meeting to discuss these issues would be useless.

According to Dr. Kevorkian's book, many other medical oncologists expressed similar views to my own. Some others admitted to performing euthanasia, or wished that it was legal. None referred any patients for Dr. Kevorkian's services.

As a guest on the Phil Donahue Show, Dr. Kevorkian said that all his efforts to attract terminally ill patients for assisted suicide had been in vain. Of 60 people who contacted him, he turned down 57 because of depression, treatable medical illness, or other reasons. Because of this experience, Dr. Kevorkian feels that euthanasia and assisted suicide should also be allowed for nonterminally ill people with unrelieved suffering.

Dr. Kevorkian appears to be a "loose cannon" in the euthanasia debate. Most pro-euthanasia advocates have distanced themselves from him. His book, *The Goodness of a Planned Death,* recounts on page after page gory details from the history of human execution, and then presents his agenda for medical experimentation on death row prisoners, painting a frightening and sickening picture.

Before embracing Dr. Kevorkian's views, you should read them in depth and determine their real nature and value for yourself. Warning—it takes a strong stomach to read his book!

FOR FURTHER INFORMATION...

A newsletter, *Cancer Pain Release,* covering cancer pain management throughout the world is available at $12/year from: *Cancer Pain Release, 610 Walnut Street, Madison, WI 53705.*

The *California Cancer Pain Coalition* is an organization devoted to the promotion of education about modern cancer pain treatment for both medical professionals and the general public, and welcomes inquiries and tax-deductible contributions at: *California Cancer Pain Coalition, 220 Ware Road, Woodside, CA 94062; 415-851-2881.*

Those wishing to help create a palliative care inpatient unit and associated consultation services for USC, UCLA, Drew/ King Medical Center, and UC Irvine may direct inquiries and tax-deductible contributions to: *Cancer Pain Service, Professional Staff Service, LAC + USC Medical Center, 1739 Griffin Ave., Los Angeles, CA 90031.*

To help defeat the California euthanasia initiative, write or call: *Californians Against the Euthanasia Initiative, Cavalier & Associates, 1121 L Street, Suite 810, Sacramento, CA; 916-4434-8060.*

References

CHAPTER 1

[1]National Hospice Organization, "Standards of a Hospice Program of Care," 1982, 1.

CHAPTER 3

[1]Anonymous, "It's over Debbie," *JAMA* **259,** No. 2, p. 272, Jan. 8, 1988.

[2]Jones E.: from *"Death, Dying and Euthanasia."* Edited by Dennis J. Horan and David Mall. Aletheia Books, University Publications of America, 403–404, 1980.

CHAPTER 4

[1]Society for the Right to Die, 250 West 57th St., New York, NY 101007, (212) 246-6973.

[2]Lasnover, AL, MD (Chairman): Report to the CMA Council from the Committee on Evolving Trends in Society Affecting Life. August 1, 1987.

CHAPTER 5

[1]Kamisar, Y, "Some Non-Religious Views Against Proposed Mercy Killing Legislation." *Minnesota Law Review* **42,** 1958.

[2]Hamel, Ron, and DuBose, Edwin: Active Euthanasia, Religion, and the Public Debate. The Park Ridge Center, pp. 45–79, 1991.

[3]General Assembly of Unitarian Universalists, *1988 Proceedings,* p. 74.

[3]Bettam. Israel: *Responsa 78,* "Euthanasia," *American Reform Responsa* **60,** pp. 107–120, 1950. This Responsa was drafted in response to a bill signed by 2000 physicians in New York State in 1948 supporting legalization of euthanasia.

[3]Muggeridge, Malcolm, "This Programme Courtesy of Lucifer Inc." *In* Roy Bonisteeel, *Searching for Man Alive.* Ontario, Totem Books, 1980.

[3]*See* Desai, Prakash: "Medical Ethics in India," *Journal of Medicine and Philosophy* **13,** pp. 231–255, Aug. 1988.

[3]Lesco, Phillip: "Euthanasia: A Buddhist Perspective," *J Religion Health* **25,** p. 55, Spring 1986

CHAPTER 6

[1]Humphry, Derek, with Wickett, Ann, *Jean's Way,* Fontana Books, New York, 1978.

[2]Humphry, Derek, *Final Exit,* Carol Publishing, Secaucus, New Jersey, 1991.

[3]Humphry, Derek, and Wickett, Ann, *The Right to Die,* Harper and Row, New York, 1987.

CHAPTER 8

[1]Lindblom U, et. al.: Pain terms—A current list with definitions and notes on usage. *Pain* (Suppl) 3, S215–S221, 1986.

[2]Bonica JJ: Treatment of Cancer Pain—Current Status and Future Needs. In: Fields HL, Dubner R, Cervero F, eds. *Advances in Pain Research and Therapy*, vol. 9. Proceedings of the Fourth World Congress on Pain. New York: Raven Press, 589–616, 1985.

[3]Levin, DN, Cleeland, CS, Dar, R, "Public Attitudes Towards Cancer Pain." *Cancer* **56,** 2337–2339, 1986.

[4]Goodwin JS, et al: Knowledge and use of placebos by house officers and nurses. *Ann. Int. Med.* **91,** 106–110, 1979.

[5]Pepper, OHP: A note on the placebo. *Ann J Pharm* **117,** 409–412, 1945.

[6]Beecher, HK: The Powerful Placebo. *JAMA* **159,** 1602–1606, 1955.

[7]Parkhouse, J: Placebo reactor. *Nature* **199,** 308, 1963.

[8]Angell, Marcia: The Quality of Mercy. *New England J Med* **306,** 98–99, 1982.

[9]Foley, KM: The Treatment of Cancer Pain. *New England J Med* **313,** 84–95, 1985.

[10]Inturrisi, CE: Management of Cancer Pain. *Cancer* **63,** 2308–2320, 1989.

CHAPTER 10

[1]Riley, G., Lubitz, J., Prahoda, R., Rabey, E.: "The Use and Cost of Medicare Services by Cause of Death," *Inquiry.* **24,** pp. 233–244. Fall 1987.

[2]National Center for Health Statistics, unpublished data, 1985.

[3]Baker, M., Kessler, L., Urban, N., and Smucker, R. "Estimating the Treatment Costs of Breast and Lung Cancer." *Medical Care* **29,** pp. 40–49, 1991.

[4]Hellinger, F. "Costs of AIDS, HIV Care." *Inquiry*, Fall 1991.

[5]Mahoney, J. (Director, National Hospice Organization) personal conversation.

[6]Levin, D.N., Cleeland, C.S., Dar, R. "Public Attitudes Toward Cancer Pain." *Cancer* **56,** 2237–2339, 1985.

APPENDIX V

[1]Quill, T. "Death and Dignity—A Case of Individualized Decision Making." *New England J Med* **324** (10), pp. 691–694, March 7, 1991.